KU-346-252

Morality and the New Genetics

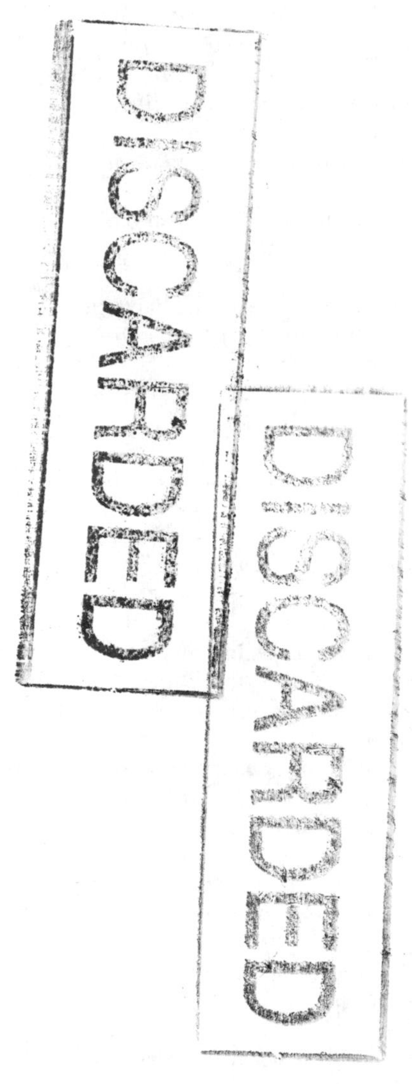

Jones and Bartlett Series in Philosophy
Robert Ginsberg, General Editor

A. J. Ayer, 1994 reissue with new introduction by Thomas Magnell, Drew University
Metaphysics and Common Sense

Francis J. Beckwith,
University of Nevada, Las Vegas, Editor
Do the Right Thing: A Philosophical Dialogue on the Moral and Social Issues of Our Time

Anne H. Bishop and John R. Scudder, Jr.,
Lynchburg College
Nursing Ethics: Therapeutic Caring Presence

Peter Caws, The George Washington University
Ethics from Experience

Joseph P. DeMarco, Cleveland State University
Moral Theory: A Contemporary Overview

Bernard Gert et al., Dartmouth College
Morality and the New Genetics

Michael Gorr, Illinois State University, and Sterling Harwood, San Jose State University, Editors
Crime and Punishment: Philosophic Explorations

Joram Graf Haber,
Bergen Community College, Interviewer
Ethics in the 90's, a 26-part Video Series

Sterling Harwood,
San Jose State University, Editor
Business as Ethical and Business as Usual: Text, Readings, and Cases

John Heil, Davidson College
First Order Logic: A Concise Introduction

Gary Jason, San Diego State University
Introduction to Logic

Brendan Minogue, Youngstown State University
Bioethics: A Committee Approach

Marilyn Moriarty, Hollins College
Writing Science through Critical Thinking

Linus Pauling, and Ikeda Daisaku,
Richard L. Gage, Translator and Editor
A Lifelong Quest for Peace, A Dialogue

Louis P. Pojman,
The University of Mississippi, and
Francis Beckwith,
University of Nevada Las Vegas, Editors
The Abortion Controversy: A Reader

Louis P. Pojman,
The University of Mississippi
Life and Death: Grappling with the Moral Dilemmas of Our Time

Louis P. Pojman,
The University of Mississippi, Editor
Life and Death: A Reader in Moral Problems

Louis P. Pojman,
The University of Mississippi, Editor
Environmental Ethics: Readings in Theory and Application

Holmes Rolston III,
Colorado State University, Editor
Biology, Ethics, and the Origins of Life

Melville Stewart, Bethel College
Philosophy of Religion: An Anthology of Contemporary Views

Dabney Townsend,
The University of Texas at Arlington, Editor
Aesthetics: Classic Readings from the Western Tradition

Robert M. Veatch,
Georgetown University, Editor
Cross-Cultural Perspectives in Medical Ethics

Robert M. Veatch,
Georgetown University, Editor
Medical Ethics, Second Edition

D.P. Verene, Emory University, Editor
Sexual Love and Western Morality, A Philosophical Anthology, Second Edition

Morality and the New Genetics

A Guide for Students and Health Care Providers

Bernard Gert, Ph.D., *Dartmouth College*
Edward M. Berger, Ph.D., *Dartmouth College*
George F. Cahill, Jr., M.D., *Harvard Medical School*
K. Danner Clouser, Ph.D., *The Pennsylvania State University*
Charles M. Culver, M.D., Ph.D., *Florida College of Physician Assistants*
John B. Moeschler, M.D., F.A.A.P., F.A.B.M.G., *Dartmouth College*
George H.S. Singer, Ph.D., *University of California at Santa Barbara*

JONES AND BARTLETT PUBLISHERS
Sudbury, Massachusetts

Boston London Singapore

QZ 50
THE LIBRARY
POSTGRADUATE CENTRE
MAIDSTONE HOSPITAL
A060447

Editorial, Sales, and Customer Service Offices
Jones and Bartlett Publishers
40 Tall Pine Drive
Sudbury, MA 01776
800-832-0034
508-443-5000

Jones and Bartlett Publishers International
Barb House, Barb Mews
London W6 7PA
UK

Copyright © 1996 by Jones and Bartlett Publishers, Inc.

All rights reserved. No part of the material protected by this copyright notice may be reproduced or utilized in any form, electronic or mechanical, including photocopying, recording, or by any information storage and retrieval system, without written permission from the copyright owner.

Library of Congress Cataloging-in-Publication Data

Morality and the new genetics : a guide for students and health care providers / Bernard Gert . . . [et al.].
p. cm. — (Jones and Bartlett series in philosophy)
Includes bibliographical references.
ISBN 0-86720-520-2 (paper)
1. Medical genetics—Moral and ethical aspects. 2. Human Genome Project—Moral and ethical aspects. I. Gert, Bernard, 1934– .
II. Series.
RB155.M67 1996
174'2--dc20 96-1781
CIP

Acquisitions Editors: Arthur C. Bartlett and Nancy E. Bartlett
Production Coordinator: Marilyn E. Rash
Senior Manufacturing Buyer: Dana L. Cerrito
Typesetting: Bookwrights
Printing and Binding: Edwards Brothers
Cover Printing: John P. Pow Company
Cover Design: Marshall Henrichs
Cover Concept: Gary Archambault

Printed in the United States of America
00 99 98 97 96 10 9 8 7 6 5 4 3 2 1

To our children and grandchildren,
who will have to live with the new genetics

Contents

Authors

Bernard Gert, Ph.D.
Eunice and Julian Cohen Professor for the Study of Ethics and Human Values, Dartmouth College
Adjunct Professor of Psychiatry, Dartmouth Medical School

Edward M. Berger, Ph.D.
Professor, Biological Sciences
Dean of Graduate Studies and Co-Director, Molecular Genetics Center, Dartmouth College
Member, Human Genome Organization (HUGO)

George F. Cahill, Jr., M.D.
Professor of Biological Sciences, Dartmouth College
Professor of Medicine, Emeritus, Harvard Medical School

K. Danner Clouser, Ph.D.
University Professor of Humanities
Department of Humanities, The Pennsylvania State University College of Medicine, The Milton S. Hershey Medical School

Charles M. Culver, M.D., Ph.D.
Program Director
Florida College of Physician Assistants, Miami

John B. Moeschler, M.D., F.A.A.P., F.A.B.M.G.
Associate Professor of Pediatrics, Dartmouth Medical School
Director, Center for Genetics and Child Development, Dartmouth–Hitchcock Medical Center
Associate Director, New Hampshire University Affiliated Program for Persons with Disabilities
Medical Director, New Hampshire Genetics Services Program

George H.S. Singer, Ph.D.
Professor, Graduate School of Education, University of California at Santa Barbara

Preface

This book is the result of three years of discussions at Dartmouth among philosophers, physicians, scientists, and specialists in dealing with the problems associated with genetic disabilities. In the summer of 1990, we received a three-year grant from the National Institutes of Health (NIH). This grant was from the Ethical, Legal, and Social Implications (ELSI) section of the Human Genome Project (HGP). Our proposal was the first grant funded through the ELSI program for work in moral philosophy and ethics. The evaluation forms accompanying the ELSI award recommended that we consider how the account of morality presented in a book by one of us could be applied to the new genetics that would result from the HGP. They also expressed interest in our using the analysis of the concept of malady developed by some of us. In particular, they stated a desire to see if and, if so, how the new genetic information that would be gained from the HGP would affect our views about what it meant to be suffering from a malady. More generally, the forms asked us to show how our philosophical analyses could be of some use in developing policies to deal with the moral issues that were already arising from the new genetics.

Nine of the chapters here, which have gone through several revisions, constitute most of the book. However, we have also included an article on genetic counseling by a very prominent Canadian counseling group as Chapter 4. We wanted to include that article because it is a paradigm case of the way that principlism, currently the most widely used method of moral reasoning in medicine, is being applied to real cases of genetic counseling. We think it is instructive to compare the usefulness of principlism with the usefulness of our account of moral reasoning. Also, the moral issues that arise in genetic counseling have become a central focus of our concern. We think that this book may help to resolve some of these issues. Unlike many of the issues concerning confidentiality, those central to the problems of insurance coverage and employment practices, the moral issues that we discuss concerning genetic counseling do not require sophisticated knowledge of our political and economic system. We have focused on problems where we believe that we have enough of the relevant knowledge so that what we say can provide real help to people facing those problems.

The plan of the book, as indicated by the table of contents, is as follows. The first chapter gives a brief history of the HGP. This chapter, together with the Appendix, is intended to introduce some of the scientific terminology that is used later in the book and to provide the kind of background knowledge of genetics that will enable any reader to understand why and how various moral problems have arisen and will continue to arise from the new genetics.

We then begin to deal with these problems. The second chapter provides what we hope is an understandable and usable account of morality. We hope that this account will enable the readers to appreciate that our common morality is quite sophisticated and to feel more confident about using their previous understanding of morality to deal with the problems that have arisen and will arise from the new genetics. The third chapter is an account of principlism, which in the past decade has become the most commonly used method for dealing with moral issues in medicine. In this chapter, we try to point out both the appeal and the weakness of this way of dealing with the difficult moral problems that will arise from the new genetics.

Chapter 4 is the article, "Ethical and Legal Dilemmas Arising during Predictive Testing for Adult-Onset Disease: The Experience of Huntington Disease" from the *American Journal of Human Genetics*, by a prominent Canadian genetic counseling group. In this article, principlism is explicitly used in order to deal with nine cases of genetics counseling for Huntington disease. Chapter 5 is an application of our account of morality to these same cases. We hope to show that, even when the answers are the same, our account of moral reasoning is more useful than the method afforded by principlism. We also believe that our account provides different and more acceptable answers to some of the more difficult cases.

In Chapter 6, we deal more generally with the problems of genetic counselors and some common misunderstandings of their goals and duties. We concentrate on the slogan of nondirectiveness in counseling, showing how it can lead to unnecessary problems. We propose facilitating a more highly developed concept of informed consent as a preferable guide for genetic counselors.

In Chapter 7, we begin our analysis of genetic malady and try to show how the new genetics may lead to some changes in what we regard as a malady and who we regard as having a malady. It may, indeed undoubtedly will, also lead to changes in the way we regard maladies and the people who suffer from them. In Chapter 8, we build on the concept of genetic malady developed in Chapter 7 and identify what we call its "morally relevant features," particularly those that determine the seriousness of the malady. These features of genetic maladies should be used to determine what kind of screening, if any, should be done with regard to those genetic maladies. We consider why it is now appropriate not to screen children for some genetic maladies, but why the discovery of new treatments may make it appropriate to screen for them.

In Chapter 9, we deal with the issue of abortion, one of the central problems of genetic counseling. However, we do not discuss this issue in the way it is normally discussed, as a conflict between the rights of the pregnant woman and the welfare of the fetus. Rather, we are concerned with abortion because of possible genetic maladies and the problems that this raises for all of those working in genetic counseling centers. We address the policies on abortion that should be adopted by these centers and how these policies should be arrived at.

In the final chapter, we examine germ-line gene therapy, a method of treatment that the new genetic discoveries may seem to make the treatment of choice for dealing with genetic maladies. In this chapter, we point out that except in very rare cases germ-line gene therapy is not necessary to prevent children from being born with genetic maladies. This means that the primary function of germ-line gene therapy will be to introduce new and "improved" traits. We argue that the risks inevitably associated with germ-line gene therapy outweigh the benefits from its adoption.

We regard the book as a unified whole, one in which the account of moral reasoning and the analysis of the concept of a genetic malady and its morally relevant features are used to deal with a selected range of problems. We have limited ourselves to those problems that we regard as not requiring the kind of reliable economic and sociological information that is not currently available. They are the kinds of problems in which philosophical analysis, if combined with knowledge of the facts about medicine, science, and genetics generally agreed on by physicians, scientists, and genetic counselors, can actually be of some practical value.

This does not mean that we offer unique solutions to the problems that we discuss, because some of these problems do not have unique solutions. It is often helpful simply to rule out unacceptable solutions and to show the range of acceptable solutions. One of the main virtues of our account of morality is that we do not say anything that, on reflection, is surprising. However, we believe that by providing an explicit and clear account of moral reasoning we may enable people to see solutions to problems that they may have missed without this account. We hope that this increased understanding of morality will help people to deal with the many important problems that we do not explicitly discuss.

We realize that there are many, supposedly conflicting, accounts of morality and that we are presenting only one of these accounts. We think these conflicting accounts are the result of philosophers attempting to revise our ordinary understanding of morality so as to arrive at unique solutions to some important moral problems. By recognizing that morality does not always provide unique solutions to moral problems, we have avoided the distortions of morality that are the inevitable result of attempting to provide such solutions. We do not think that we rule out as a morally acceptable solution to a moral problem any solution that is seriously proposed by those holding any other moral theory, including all plausible forms of consequentialism, deontology, virtue theory, casuistry, feminism, or principlism. We do deny that some of the solutions presented by these theories are the only morally acceptable solutions. Unlike that of most other moral theories, our goal is not to resolve moral controversies, but rather to provide a framework for moral reasoning that all parties to a controversy can accept, thus allowing for fruitful discussion of the problems involved.

Although we have assembled the chapters in their present order because we believe that this arrangement emphasizes the systematic character of the book, we are aware that some may prefer to read it in a different order. They may prefer to follow the order of discovery rather than the order of presentation. In that case, we recommend starting with Chapters 4 and 5, which present alternative ways of deciding how to deal with nine genetic counseling cases. It was our experience in dealing with those cases that convinced us of the need to present an account of morality that was more useful than that presented by principlism. Similarly, one may prefer to read the chapters on abortion and gene therapy to see how we deal with these specific problems before reading the chapters that provide the foundation for our way of dealing with them.

Although *Morality and the New Genetics* is a truly collaborative enterprise, individual members of the group took primary responsibility for writing each chapter. A draft of each chapter was presented to the group at a meeting in which the rest of the group raised questions, offered suggestions, and otherwise caused problems for the author. Second drafts of most of the chapters also were presented to the group; in addition, there were discussions of the chapters among individual members of the group.

The principal author of the Chapter 1 is George Cahill. George was intimately involved with all of the major parties in the development of HGP. The group discussed this chapter and offered suggestions, but this chapter, more than any of the others, is primarily the work of one author. The principal author of the Chapter 2 is Bernard Gert. Although this chapter is primarily a focused summary of the account of morality that he presents in his book *Morality: A New Justification of the Moral Rules* (Oxford University Press, 1988), it is far more understandable to nonphilosophers because of the many suggestions from the group.

Dan Clouser was primarily responsible for Chapter 3, the account and criticism of principlism. This chapter is a development of previous work by Clouser and Gert, especially "A Critique of Principlism" (*The Journal of Medicine and Philosophy,* Vol. 15, No. 2, April 1990, pp. 219–236), and "Morality versus Principlism" (*Principles of Health Care Ethics,* edited by Raanan Gillon, John Wiley and Sons, 1994, pp. 251–266). The present chapter serves as an introduction to the next two chapters, which provide alternative methods for dealing with ethical decisions that must be made in actual cases in genetic counseling.

The Canadian article, "Ethical and Legal Dilemmas Arising during Predictive Testing for Adult-Onset Disease: The Experience of Huntington Disease" (Chapter 4), is included because it is an excellent example of principlism in action. Our group met to discuss the dilemmas presented by the nine cases and the solutions to them put forward in this article. We considered the article very important and extremely well done. However, recognition of its inadequacies persuaded the group that there was a need for a more useful method for dealing with the problems raised by the new genetics.

Chapter 5, for which Bernard Gert was primarily responsible, is an attempt to provide a more useful procedure for dealing with the dilemmas that emerged in the nine cases presented in the Canadian article. This chapter is the best example of the group working together because it arose from the meetings in which the group discussed the Canadian article and tried to apply Gert's account of morality to these cases. In a sense, this chapter is the heart of the book because not only is it the first chapter that resulted from discussion by the group, but it is the chapter that persuaded the group that we could indeed say something useful for dealing with the problems raised by the new genetics.

Chapter 6, whose principal author is George Singer, involved extensive collaboration between Singer and Gert. It is primarily concerned with genetic counselors, showing the problems caused by their adoption of the slogan of nondirective counseling and offering an alternative model based on facilitating informed consent. This chapter shows the usefulness of the distinction between the moral rules and the moral ideals and shows how easy it is to misuse principlism, especially the principle of autonomy.

Chapter 7, for which Chuck Culver had primary responsibility, is an attempt to deal with the problems that will arise from the new concept of a genetic malady. This chapter is built explicitly on the foundation provided by an earlier article, "Malady: A New Treatment of Disease" (*The Hastings Center Report*, Vol. 11, No. 3, June 1981, pp. 29–37) by Clouser, Culver, and Gert on the concept of a malady.

Chapter 8, whose principal author is Ed Berger, is an attempt to set up a classification scheme of genetic maladies that will be helpful to those who will have to decide which genetic maladies should be tested for and that will aid genetic counselors when counseling about the new genetic maladies that will be discovered. This chapter was one of those most affected by group discussion, and several different versions were presented to the group before arriving at the chapter presented here.

John Moeschler is the principal author of Chapter 9, which concerns abortion for genetic maladies but concentrates on the problems that this causes for those who counsel and perform the abortions; it is not a standard treatment of the pros and cons of abortion. It underwent many revisions and entailed extensive collaboration between Moeschler and Gert.

The final chapter, whose principal author is Ed Berger, deals with germ-line gene therapy and was developed from an article, "Genetic Disorders and the Ethical Status of Germ-line Gene Therapy" (*The Journal of Medicine and Philosophy,* Vol. 16, No. 6, December 1991, pp. 667–683), that Ed Berger and Bernard Gert wrote several years ago.

As stated before, this book was developed by means of monthly seminars in which we discussed moral theory, genetics, and the relation between the two. George Cahill, Ed Berger, John Moeschler, and Bernard Gert attended these seminars for all three years of the grant; Chuck Culver attended the seminars for the first half of the grant; and George Singer attended the seminars for the second half. Dan Clouser, the only member of the group whose primary position was not at Dartmouth, attended the seminars during the summers when he was in Hanover and while serving as a Visiting Professor during the spring term of 1993. Notes on the meetings were kept by Barbara Hillinger, Assistant to the Director of the Dartmouth Ethics Institute. She was also responsible for coordinating the meetings and arranging for the distribution of readings; she has been indispensable in preparing this manuscript for publication. Michael Kligman, an ethics and psychiatric fellow, attended the seminars for almost two years. Attending even longer was Binette Elliott, who wrote her Master of Arts and Liberal Studies thesis, "Ethical Problems in Predictive DNA-Based Genetic Testing Created by the Artifact of Health Insurance," during this same period. Both made significant contributions to the discussions at the seminars. Several undergraduate students participated for one year, and several visitors attended the seminars and took part in the discussions.

Acknowledgments

We distributed earlier versions of the manuscript to several people who have provided us with valuable comments. One of the earliest and most important sets of comments came from Lois Banta, a biologist at Haverford College. Other early and valuable sets of comments came from Robin Gregg and Dorothy Wertz of the Shriver Center for Mental Retardation, Inc. Earlier versions of the manuscript were used as a text in an undergraduate philosophy of medicine course at Dartmouth, and many useful suggestions were incorporated owing to the comments of the students in those classes.

We are grateful for the grant from NIH, which enabled us to spend the time that was necessary to produce a truly collaborative work. We hope that this book will enable others to benefit from that collaboration. As Principal Investigator for the Project, I want to thank all who participated in it. I could not have asked for more cooperation from the authors of the various chapters. Their willingness to revise their chapters, even after the grant had officially ended, showed their commitment to the project. *Morality and the New Genetics* is truly an interdisciplinary accomplishment; not only could it not have been written by any single person, but it could not have been written by a group of people from a single field or discipline. We hope that the reader will benefit as much from our efforts as we have.

The Institute for the Study of Applied and Professional Ethics at Dartmouth was indispensable to the successful completion of this book. They not only helped to prepare the initial grant proposal to NIH, but assisted me with all of the administrative details that I was incapable of handling. In particular, without Barbara Hillinger, Assistant to the Director of the Dartmouth Ethics Institute, this book would never have been completed. I have also benefited from funds provided by Eunice and Julia Cohen, who not only endowed the Eunice and Julian Cohen Professorship for the Study of Ethics and Human Values at Dartmouth, but also provided additional funds to support the work of the holder of that chair. That support enabled me to continue working on this project after the grant ended. This book is one tangible result of their generosity.

Bernard Gert

1

A Brief History of the Human Genome Project

This chapter summarizes human genetics and its history with simple descriptions of modes of inheritance, using the commonly used terms from the genetic literature. It also describes current efforts to create genetic maps and to sequence the 3 billion bases in the human genome. An Appendix at the back of this book summarizes the genetic principles of inheritance in human beings.

Antiquity

Awareness of the inheritance of both human appearance and behavioral characteristics dates from ancient times. In a similar vein, selective breeding of domestic animals and plants, both for greater yields and for ease of management, produced strains that maintained these better characteristics in subsequent generations. The evolution of today's corn from the minuscule wild maize of pre-Colombian America is one excellent example. Others include the draft horse, both beef and dairy cattle, and especially the various breeds of dogs for different uses such as herding, retrieving, pointing, or simply companionship. These examples represent what can be done with intentional genetic manipulation, employing appropriate selection for physical and behavioral traits.

Emerging Concepts

Although many characteristics such as body height and build, hair and skin color, and some behavioral traits were recognized in human beings as familial, it was the inheritance of human abnormalities that led to a more definitive science. In India two millennia ago, Susrut described diabetes to be both "passed through the seed" (genetic) as well as influenced by environmental factors ("an indolent spirit and love of rich foods and sweet wines"). More scientific approaches began 200 years ago when the French physician Maupertuis recognized that polydactyly (more than five fingers) was passed from one generation to the next (McKusick 1994). If one parent, either father or mother, expressed the abnormality, about half of the children of either sex would be similarly affected. Today we would call this mode of inheritance *autosomal dominant* (see Appendix, Figure 5).

Similarly, hemophilia and color blindness had both been noted to be familial but expressed only in males, and a number of papers in the early nineteenth century described the unique route of maternal inheritance with half of uncles on the mother's side expressing the disease. Likewise, the sisters of the affected boy might or might not pass the trait to their sons. Ancient Hebrew law even stated that circumcision could be obviated in a male infant either if his brother exhibited excessive bleeding or if the same tendency had been noted in a maternal uncle. Thus, "sex linked" inheritance (see Appendix, Figure 7) as we know it today was recognized in antiquity. Horner, a British physician, observed in 1876 that color blindness is inherited in boys from mothers whose brothers had the problem, another example of a classical sex-linked genetic history.

Another British physician, John Adams, also in the early nineteenth century, made a more imaginative and incisive genetic observation when he realized that certain traits (or diseases) could come through the families of apparently normal parents (*phenotypically* normal but *genotypically* not normal, in genetic language). But these abnormal traits were manifested in only a small proportion of the children. Even more striking, this phenomenon occurred more frequently in kindreds in which the parents were not only normal, but blood relatives (*consanguinity*). He had brilliantly described recessive inheritance (see Appendix, Figure 6) in the early 1800s, a full century before the recognition of Mendel's great contributions to our understanding of inheritance (recognized independently by three European scientists some 30 years after Mendel's publication).

Mendel and Pea Counting

Thus the genetic foundation for inherited disease in human beings had been laid without knowledge of genes, chromosomes, DNA, and the like. It was the mathematician-monk Mendel, in his work on breeding garden peas, who put genetics into a quantitative discipline through the inheritance of "factors" that determine the physical features, such as plant height and flower color, in subsequent generations. Mendel correctly postulated that two copies of each factor are present in each of the parents and only one copy of each factor in the sex products—the gametes, or egg and sperm (pollen in plants), respectively (see Appendix, Figure 2). Mendel also pointed out the random or independent inheritance of one pair of factors—say, pea or flower color—with another pair, "wrinkled" versus "smooth," and that certain factors were dominant over the other member of a pair—a pea with one yellow factor and one green factor would be yellow in color (a yellow phenotype but a mixed genotype) and indistinguishable from a pea with both yellow factors (yellow phenotype and genotype). The latter, however, would produce only yellow peas in subsequent generations and the former, mixtures of the two colors.

At the turn of the twentieth century, Mendel's work, then four decades old, was discovered by several Europeans studying heredity and was publicized throughout the world's scientific community. Subsequently, Sutton, in 1904, attributed to the small stained organelles in the nucleus or in the nuclear remnant of dividing cells, the *chromosomes,* the role of bearing the inherited traits. The British physician Archibald Garrod characterized certain rare human diseases as being due to loss of a "Mendel factor," presaging by decades the roles of genes in encoding proteins, especially enzymes in metabolic pathways. Several years later, in 1909, Johannsen called these inherited factors "genes," providing the science of genetics with its name.

Thus, by 1910, genetics was off and running as a creditable and exciting science with Thomas Hunt Morgan using the common fruit fly, *Drosophila melanogaster,* as a model to study the inheritance of genetic traits. Much of what we know of genetics today stems from these classical experiments of Morgan and his pupils. In 1911, the X chromosome was shown to be related to sexuality, 2 X's being female, and a single X with its counterpart, Y, being male in both human beings and the fruit fly. E. B. Wilson attributed the sex-linked genes responsible for hemophilia and color blindness in human beings to be located on the X chromosome, similar to the many X-linked factors being described by the Morgan group in flies.

Early Politics and Ethical Issues

The enthusiasm for the new science led to two movements, both leading to bad results: one for human well-being and the other for scientific progress. Because inheritance was deemed so important and was placed on a strong scientific basis by both Darwin and Mendel and their followers, the eugenic movement (probably better termed a cult) concluded that the future human population should be improved by selective breeding and culling. Kevles (1985) wrote an extensive treatise on this topic. This cult movement is an excellent example of how the conviction that one is attempting to achieve a good objective often leads one to adopt morally unacceptable means. The second unfortunate result was the attempt to quantify all human characteristics by the simple mathematical maneuvers of Mendel. Thus arose a computational school of genetic science with many of the world's great minds—Haldane, Huxley, Hogben, Galton, Jennings—attempting to calculate both population and individual characteristics, including physical and behavioral traits, in terms of simple, single-gene models. This period of "softness" in genetic science even abetted the eugenicists in their naive attempt to improve the human gene pool. The final result of these activities was the Nazis' attempt to select for racial purity with improved physical and emotional characteristics. This early history of eugenics shows how easy it is for a worthy goal to be perverted and how common it is for even great scientists to overestimate their scientific knowledge. We are now in another period of time when some scientists think that they know enough about genetics to attempt to achieve the same worthy goal of improving the future human population. It is one of the major goals of this book to provide such a clear account of morality that it will be extremely difficult for this goal to be used as a justification for using morally unacceptable means.

Linkage

Hogben did, however, suggest with his mathematical approach that genes situated close together on the same chromosome could be used for locating the anatomic position (locus) of other genes on that chromosome—in other words, gene "mapping" by genetic linkage—and this was in the early 1930s. Geneticists studying mice, fruit flies, and the bread mold, *Neurospora*, were cataloguing numerous traits (factors or, more properly, genes) and found that variations in a single gene (an *allele*) would frequently be inherited with a specific variation in another gene (a particular allele of this second gene). Thus these two genes must be close together or "linked" (see Appendix, Figure 3). Mice were found to have twenty large "linkage" groups, matching the number of chromosomes in mice. Soon thereafter, Haldane's group calculated how close together, in genetic terms, were hemophilia and color blindness on the human X-chromosome, because over many generations the two would occasionally separate owing to crossing-over (recombination) in gamete (egg or sperm) production (see Appendix, Figure 4). Thus the concepts of linkage and its mathematical measurements were firmly established.

Enter DNA

Keep in mind that all of the foregoing was "paper" biology centered on breeding experiments in human beings, animal models, and plants. Little was known about the underlying structure and function of the genetic information. The author of this chapter was taught (in 1944) in college biology that genes were located in the protein–nucleic acid complexes in the "forty-eight" (!) chromosomes in human beings with the genetic specificity for the code probably residing in the protein part because it contained twenty different building blocks (the amino acids) compared with the paltry four building blocks in nucleic acids (A, G, C, and T). The latter supposedly provided some kind of order or structural support. At this same time, Avery and colleagues at the Rockefeller Institute were finding that one strain of pneumococcus could be made into another strain simply by transferring the nucleic acid component (DNA) free of all protein. It still took a number of years, however, before this most fundamental contribution, that DNA is the vehicle for genetic information, was entered into the standard biological texts.

Mapping

Mapping, or, more properly, assignment of human genes to locations on various chromosomes, proceeded very slowly, because controlled breeding experiments in human beings could not and cannot be done. Mohr noted in 1951 that a gene for a blood protein factor, Lutheran, in some families appeared to be associated with another gene controlling salivary secretion of a red blood cell protein. In other families, it was not associated with the secretion gene. Thus *secretor* and *Lutheran* appeared linked and presumably close together on the same chromosome. Both are now assigned to chromosome 19, physically corroborating the earlier family linkage studies of Mohr.

The major contribution of Watson and Crick came in 1953, the description of the molecular structure of the genetic material—the double helix. More important was the emphasis on the specific pairing of the two sets of nucleic acids, A to T and G to C (see Appendix, Figure 1). This pairing is the basis of all nature because it is central not only to the accurate duplication of genes in cell growth and reproduction, but also to how genes are made into specific messages for synthesis of specific proteins.

This pairing of the four bases in DNA, A to T and G to C, is the basis of today's molecular science. A and T bind and fit like a key in a lock, and the same for G and C. A chromosome is simply an immense single molecule of alternating sugar and phosphate molecules with one of the four bases attached to each molecule of sugar (see Appendix, Figure 1). Thus, if the sequence -T-A-G-G-C-T-A-G-G-C- is located somewhere in a given chromosome, its complementary strand in the double helix will contain -A-T-C-C-G-A-T-C-C-G-. If a chemist were to make the latter sequence, label it by adding a colored tag or a radioactive element (creating a *probe*), and add it to a test tube full of DNA, the labeled sequence would search out and bind to its complement wherever it appeared in the millions to billions of DNA sequences in the test tube. The labeled probe could even be added to a microscopic slide containing the DNA to be examined and anatomically identified as to where it binds—for example, to which chromosome and even to where on that chromosome. By random chance alone, these 10 bases would be expected to find their complementary counterpart once in every 4^{10} stretches of 10 bases or once per 1,048,576 bases. This concept is fundamental to the locating ("mapping") of genes and nongene sequences throughout the genome.

Clinical Genetics as a Medical Specialty

Human genetics was, by today's standard, proceeding at a snail's pace. Tijo and Levan in 1956 found that human beings had forty-six and not forty-eight chromosomes and, several years later, Lejeune noted a forty-seventh chromosome in persons with Down syndrome. Other significant genetic concepts were formulated at about that time; for example, Allison, in England, made the provocative suggestion of a possible protection induced by the sickle cell trait against malaria. A trait, thereby, can be selected for in a certain environment where there is sufficient pressure for its positive effects over its deleterious effects.

Another notable contribution was made by Barr who found that female cells bear a blob of chromatin in their nuclei. It was later shown by a British mouse geneticist, Mary Lyon, that females inactivate early in embryonic life one of their two X chromosomes, the inactive one persisting as the "Barr" body. This was used for sex identification of athletes until recently; the test contained errors, which diminished its usefulness. Today's testing is done with molecular probes for sequences specific to the Y chromosome. Thus cells in both males and females have but one functioning X chromosome. In the female, however, the inactivation is random, and therefore the female carrier is usually phenotypically normal with half her cells bearing enough genetic material to do the job.

By the 1960s, genetics had become a medical subspecialty and, in 1966, Victor McKusick at Johns Hopkins Medical School published the first edition of *Mendelian Inheritance in Man,* containing 1,500 inherited characteristics, or, in proper terms, "mendelizing phenotypes." Again, the *phenotype* is the physical expression of a gene, whereas the *genotype* refers to the genetic material of that gene. Of the 1,500 characteristics, 574 had been clearly identified. In the eleventh edition of 1994, now two volumes, there are 6,678 entries: 4,458 autosomal dominant, 1,730 recessive, 412 X-linked, 19 Y-linked, and 59 located in mitochondria (McKusick 1994).

Although linkages had been found, especially on the X chromosome as exemplified by sex-linked inheritance of hemophilia and color blindness, it wasn't until 1968 that an autosomal (meaning not sex-linked) assignment of linkage was made by Donahue, an associate of McKusick. He observed a peculiar, microscopically visible stretch of chromatin on his own largest chromosome (chromosome 1). Looking at a number of blood factors, Donahue found that, in both himself and his relatives, an allele of the Duffy blood factor was linked to this observable physical change in their first chromosome. Thus *Duffy* was assigned to chromosome 1.

Technologic Advances

A major contribution to mapping and linkage of genes was subsequently made by Weiss and Green, who joined hamster and human cells into hybrid cells in tissue culture. In subsequent generations, the human chromosomes were lost, leaving a hamster cell line with, at times, a single human chromosome. When specific markers for human genes such as molecular probes for sequences in that gene or for their protein products were used, the nearby genes on that remaining human chromosome could be identified and assigned to that chromosome. Thus linkage was more easily established. This kind of mapping is termed *physical mapping* because it entails the use of chemical technology to physically locate a stretch of DNA. The other kind of mapping—genetic, or "Mendelian," mapping—requires large-scale breeding experiments or, in human beings, large pedigrees to calculate by linkage with other known genes or identified stretches of DNA where the unknown gene is located.

Another major contribution was made by Caspersson, who designed a greatly improved method for staining chromosomes, producing bands of differing intensity depending on the relative densities of G-C versus A-T in that region. Thus not only could gross abnormalities be more easily recognized, but the experienced eye could identify each human chromosome and the specific regions on that chromosome as recognized by their banding appearance. Nevertheless, by the mid-1970s, only several hundred human genes had been mapped—significant progress, but slow, given that the human genome contains perhaps 100,000 genes.

DNA and Genetics Unite

At the first human gene mapping workshop held in New Haven in 1973, new data on gene mapping were sparse, as exemplified by no genes having yet been assigned to chromosome 3, 8, or 9. Medical usage of the data was also infrequent. Nevertheless, Ruddle at Yale, with shrewd foresight, began his computer compilation of the information in anticipation of the oncoming plethora of data as the science matured. Kan and colleagues in San Francisco used molecular hybridization (a fancy term for A-T and G-C binding between a stretch of DNA and its complementary counterpart as a probe) to identify losses of globin genes in a relatively common type of anemia in human beings, thalassemia (Mediterranean, or Cooley, anemia). Two years later, in a very significant observation, Kan and Dozy (1978) noted that near the sickle cell gene was a stretch of DNA that varied among most blacks and could be used as a marker in a given family for linkage with the sickle cell gene. This variant (properly termed a *polymorphism*) served as the first use of these common molecular differences between one person and another as a method both for diagnosis (by being linked to a disease gene) and subsequently for general mapping over the entire genome. The scientific community was then prepared for a quantum leap in molecular genetics and the human being was becoming the principal model.

Large-Scale Mapping

Several events occurred in 1979: the linkage diagnosis of sickle cell anemia just described was published; and, in similar studies in England, Solomon and Bodmer suggested that the same technology could be applied to other genetic diseases and traits. In other words, unique variants in the genome sequence could be used as markers for nearby abnormal genes. At a small, rather informal conference held in 1978 in Alta, Utah, several researchers led by Botstein suggested that the use of these variants spread out over the human genome could predict in given kindreds the location of many genes by their genetic linkage to the variant. Botstein, Skolnick, Davis, and White published this suggestion in 1980 and it is now a milestone in genetics. White, in the fall of 1979, and his research fellow, Wyman, at the University of Massachusetts Medical School, isolated their first useful anonymous probe, meaning a variable stretch of DNA with no known function. This was mapped several years later to chromosome 14. This probe showed many variations among individuals (highly polymorphic and therefore very informative in genetic lingo). Subsequently, it and many others could be used as "road markers" in the overall effort to find many milestones up and down all of the twenty-three human chromosomes. White subsequently moved to Utah for better access to the large Mormon kindreds assembled by Skolnick, of which the four grandparents, both parents, and many children (six or more and, in a few families, more than a dozen) could be used for linkage analysis of polymorphisms and disease-related genes. Thus was initiated the detailed mapping of the human genome.

This started the ball rolling for a number of laboratories to search for variations, called RFLPs (restriction fragment length polymorphisms), and their linkage to other known loci and to disease entities. Within the next few years about one-third of the useful and available probes were generated by White's large laboratory in Salt Lake City, about one-third by a commercial concern—Collaborative Research in Boston, directed by Helen Donis-Keller and her team—and the remainder by a number of laboratories elsewhere in the world.

Nobel laureate Jean Dausset assembled a team of physicians and molecular biologists in Paris (CEPH for Centre Etudé Polymorphisme Humaine) who prepared samples of DNA from originally forty very large families with three living generations, including twenty-seven families from White's Utah Mormon collection, the immense Venezuela pedigree assembled by Nancy Wexler for her studies on Huntington disease, ten French families, and two others from the Pennsylvania Amish genetic studies by the Johns Hopkins group led by McKusick. These DNA pools from six hundred people have been made available by CEPH for investigators throughout the world mapping genes for diseases and other genetic loci of their interest, sparing the need for them to assemble similarly large "control" kindreds for linkage analysis.

Over the next few years, the locus of Huntington disease (chromosome 4, Gusella et al. 1983), adult polycystic renal disease (chromosome 16, Reeders et al. 1985), muscular dystrophy (chromosome X, Koenig et al. 1987), and cystic fibrosis (chromosome 7, by a large collaborative effort between Collins et al. 1993; Tsui et al. 1989; Riordan et al. 1989) were all mapped using this technology. Subsequently the product of the isolated defective genes for muscular dystrophy and for cystic fibrosis were characterized by "reverse genetics." This means finding a stretch of DNA known to bear the disease by linkage analysis, then finding the gene in the stretch, and finally determining what the gene does and where it is physically expressed—in what tissue and where in the cell. The gene for Huntington disease has just been found by this process, and it is hoped that the precise role of this gene in causing the pathology can now be characterized, using that knowledge for an eventual prevention or cure. Mapping genes, normal and diseased, was off and running.

Let's Even Consider Sequencing

In May 1985, a meeting was convened in Santa Cruz, California, by Robert Sinsheimer, a distinguished molecular biologist and also a very senior administrator at the University of California. A number of biologists active in genetics and gene mapping were present, including Nobelist Walter Gilbert. The dramatic proposition was made that the entire human genome should not only be mapped with scattered but specific road-markers along the highway, but also sequenced, meaning determining the precise order of each A, G, C, and T. A single copy (the normal has two copies each of twenty-three chromosomes, one copy from each parent) contains approximately 3 billion bases (A, G, C, or T). At the time, analysis of a single base cost more than ten dollars and it took a good scientist a day to sequence from 50 to 100 bases. Thus the suggestion to sequence the genome was not only imaginative and ambitious, but prohibitively expensive.

It should be mentioned that, by 1991, the technology had improved sufficiently to allow as many as 10,000 bases to be determined in a single day at about a dollar a base (Hunkapiller et al. 1991). In 1993, sequencing was being done by several laboratories using robotic analyzers at 500,000 bases daily, costing from ten to fifteen cents per base. Mapping, using primarily the RFLP technology by 1985, as summarized in the publication of Ruddle's computerized resource in New Haven, HGML (Human Gene Mapping Library), listed more than 1,200 locations for both genes and anonymous probes. By the 10th International Conference held in Paris in 1989, there were listed 5,510 locations, 8,450 probes including data on just under 2,000 RFLPs. The science was moving rapidly. But sequencing and mapping are greatly differing tasks in size, expense, and complexity, as the following will attest. It should be stated that very little attention was paid at that time to any moral or ethical issues that might ensue. The potential benefits for cancer, genetic diagnosis in pregnancy, and the genetic prediction for diseases in midlife and later, and even the proper prevention of genetic problems by gene manipulations were all so attractive that considerations of the privacy of the data, their possible prejudicial use in insurance and employment, and many other problematic applications were essentially obscured by the enthusiasm of the protagonists (Tauber and Sarkar 1992; Muller-Hill 1993).

Enter the Department of Energy

The Department of Energy (DOE) had a number of laboratories (residues of the atomic bomb project and subsequently nuclear energy for power production) dealing with the biological effects of irradiation, located at Oak Ridge, Los Alamos, Livermore, and Berkeley. Using their expertise in engineering technology derived from their massive projects in atomic energy and its potential dangers, they had already made several major contributions toward the genome project. One of these was the capacity to physically separate chromosomes by their size and staining, thus superseding the complex biological separation using human-hamster cell hybrids previously developed for linkage and mapping. Linkage was therefore greatly simplified because, instead of dealing with all twenty-three pairs of chromosomes, one had either single chromosome collections or pools of several chromosomes in separated samples. The field moved forward at a greatly accelerated pace (McKusick 1989a).

There were also politics. The studies of Hiroshima survivors fortunately failed to show a major genetic impact of atomic irradiation, although there were significant increases in malignancy rates in the survivors (Lindee 1995). What therefore should DOE do with all its assembled biologists and physicists in the national labs? One major contribution had already been started—namely, GenBank, a computerized repository for the primary DNA sequences of genetic material from all organisms reported in the literature. This was located at Los Alamos with a collaborating counterpart at the European Molecular Biology Laboratories (EMBL) in Heidelberg, Germany. By 1991, some 60 million bases had been recorded, about half being human and the remainder from bacteria, mice, fruit flies, and so forth. This number doubled in the subsequent two years to well over 100 million bases.

A biophysicist and administrator in DOE, Charles DeLisi, convened a meeting in Santa Fe in early 1986 to spell out the role of the Department in sequencing the entire human genome, following up on the previous meeting at Santa Cruz. A similar meeting was subsequently convened a month later by James Watson at Cold Spring Harbor. Already there was rising opposition from a number of quarters, calling the task a political boondoggle, a task that would drain funds from other more needy science or a task that would be scientifically not very productive, because a major part of the genome is made up not of genes but of "filler" DNA, or, more simply, "junk." Finally, there also began a concern with the moral problems that might result from learning too much about a person's genome and particularly the eventual use of that information. More will be said on this subsequently; in fact, it is the purpose of this book. Also, animosity arose between DOE and other federal agencies. Scientists, both federal and in the private sector, aligned either strongly for or strongly against the project. Nobel laureate Renato Delbucco published a strongly protagonistic article in *Science,* stating that knowledge of cancer, perhaps even its potential cure, would result from genome sequence information. Other leading scientists (Davis, Leder, Rechstein) took the opposing side and publicized their objections accordingly. This debate between prominent scientists added even more concern to those who were raising the ethical and moral issues, both for and against the genome project.

The Howard Hughes Medical Institute, whose asset, Hughes Aircraft, had just been sold to General Motors for more than 5 billion dollars, had been supporting not only the Utah operation of Ray White and several other mappers, but also Ruddle's data base, the Human Gene Mapping

Library. The Institute, in an ecumenical attempt, convened authorities from both DOE and NIH and invited a number of foreign representatives. The meeting, held in Bethesda, was chaired by Sir Walter Bodmer of England, a major contributor to knowledge of human genetics and immunology. A number of advances were presented, including advances in sequencing technology, techniques for isolating large pieces of DNA using pulsed field gels, mapping techniques used to connect contiguous pieces of DNA, and many others. The meeting did get the attention of the public and, more importantly, of a number of government agencies and both the legislative and the executive branches. The genome was now launched into the political sphere and into the public and scientific media. Robert Cook-Deegan has summarized these events in a volume covering the history of the genome between 1986 and 1990 (Cook-Deegan 1994).

DOE and NIH Take the Lead

Over the next year, a number of other meetings were held by DOE and NIH, as were independent hearings by the Congressional Office of Technology Assessment (OTA) and by the National Academy of Science. These activities and their published reports all culminated in 1988 in the establishment of the Genome Office at NIH, directed by Nobelist James Watson. This subsequently became the National Center for Human Genome Research (October 1989) by Congressional authorization. Advisory boards were created to serve both agencies (NIH and DOE). Meetings have been held twice annually on succeeding days because several committees are "joint," especially the one on data (Joint Informatics Task Force). Also initiated was the NIH-DOE ELSI (Ethical, Legal, and Social Issues) Working Group. On the recommendation of James Watson, 3% of the total funds was to be devoted to these activities, a rather bold and innovative move both by Watson himself and by NIH. (This book is supported by a three-year grant awarded by the National Center and the first to have a philosopher as Principal Investigator.)

Other Species

Although the genome task is labeled "Human," other species are included for a number of reasons. Bacteria, particularly *E. coli,* have served as the original models for molecular biology, such as how genes are turned on or turned off; thus the study of bacterial genetics is directly relevant to human disease. By 1995, the entire 1,830,137 base pairs of the bacterium *Haemophilus influenzae* RD, had been sequenced and its genes mapped by a large team headed by J. Craig Venter at the Institute for Genomic Research (TIGR) in Gaithersburg, MD (Fleischmann et al, 1995). A tiny worm, *C. elegans*, has less than 1,000 cells in its total body and has served as a superb model for how an animal's genome programs itself from a single fertilized cell into a multicelled adult. Its generation time is only 3.5 days and it has but 100 million base pairs in its genome compared with the 3 billion in human beings. The fruit fly is also included; it is the historic model for genetics thanks to its ease in breeding and the very large amount of knowledge already accumulated starting with Morgan's studies eight decades ago. Other relevant forms such as yeast and subhuman primates are included, but most important is the mouse (*vide infra*).

Genes are evolutionarily conserved. If a gene works well in bacteria and plants, it is probably similar in both structure and sequence in animals. Thus certain genes in mice and human beings are almost chemically identical, and probes used to identify a gene in one ofttimes react with the homologous gene in the other. Of interest, however, is that the order of the genes on the chromosomes is also usually conserved. Thus, if there is linkage between two genes in the mouse, the linkage between the two is probably also in human beings. Luckily, it is morally acceptable to breed mice, because mice are easier to breed and study compared with human beings, mapping is far easier in mice, and these data can then be extrapolated to human beings. For example, the seventeenth chromosome in mice is homologous in large part to the eleventh in human beings and, of the thirty-five mapped loci in both organisms on these chromosomes, all but two are ordered into the same sequence. Another major reason for interest in the mouse has been the recently developed technology, primarily by Capecchi of the University of Utah and Smithies, now at the University of North Carolina, that creates mutations or deletions of specific genes in the mouse and therefore produces genetic models for disease. Thus knowledge of mouse genes, of their locations, and of their sequences is directly relevant to the human genome project. Another major reason for the interest in the mouse is that many diseases are polygenic and result from the interaction of a number of inherited factors. Hypertension, coronary artery disease, juvenile and most maturity-onset diabetes, and most of the other common human diseases are polygenic. Sorting out the relative roles of these multiple genes necessitates both breeding and molecular genetics, and obviously this can be done only in the mouse.

Finally, there is also now a great flurry to map and eventually to sequence genes in plants and in domestic animals for both agricultural and pharmaceutical purposes. The cow genome has just been partially mapped with probes for each of its thirty pairs of chromosomes. New strains of vegetables resistant to pathogens as well as to frost are being developed. Human proteins are now being produced by human genes functioning in domestic animals for subsequent harvest and clinical usage without fear of human-infecting viruses such as HIV.

A few words must be said about baker's yeast (*Saccharomyces cerevisiae,* to be more proper). Maynard Olson and colleagues in St. Louis have been successful among others in inserting fragments of human genes into yeast. The yeast are then grown and examined for which sets of genes are together in a given fragment. This technology has greatly simplified mapping of genes to specific locations on specific chromosomes.

Mapping and Data Management

With all the aforementioned biological and political history, we should examine the present status of the genome project in the United States and other nations where significant programs are in operation. Once the movement for mapping the human genome began to accelerate in the late '70s and early '80s, there arose a number of needs for coordination, communication, and collaboration at all levels. A map with markers at approximately 10 centimorgans would necessitate from approximately 300 to 400 evenly spaced probes, with each having a high level of polymorphism so that they could be used for linkage studies in pedigrees. A centimorgan (cM) approximates 1 million bases, although in the Mendelian map derived from genetic linkage studies (see Appendix, Figures 3 and 4) this number of bases is highly variable, and where there is a high degree of recombination, may be as few as 10,000 to 20,000 bases, and, where recombination is infrequent, perhaps several million bases. More problematic is that, at any given location, recombination may be more or less frequent between oogenesis and spermatogenesis, meaning that the linkage map may be larger or smaller than the anatomic (physical) map at any location between male and female. There are really, therefore, two genetic maps, one for males and one for females. More simply stated, because the "genetic" distance between two genes is measured in centimorgans (% frequency of recombination between two points), the probability that they will separate in the formation of an egg or sperm, say, if 10 centimorgans apart is 10 in 100. In egg formation, this may be 20 million bases between two genes whereas, in sperm formation, perhaps 5 million. Thus the genetic map at this location would be four times larger for oogenesis whereas the physical map is the same for both processes. In other areas, the male distance may be larger. The mechanisms for this difference remain obscure.

Several data bases for biologic information were already in operation at this time, including the Protein Data Base for molecular structure at Brookhaven National Laboratories, the Protein Information Resource started by Dr. Margaret Dayhoff located at Georgetown University for the storage of amino acid sequences in proteins, and GenBank for the storage of DNA sequences, located in Los Alamos at the National Laboratories of the Department of Energy. GenBank is linked administratively, scientifically, and literally by computer to a similar facility, EMBL, in Heidelberg, Germany. As mentioned previously, the Human Gene Mapping Library was started by Frank Ruddle in New Haven for the storing of human genetic information in a number of data bases accessible to scientists by computer over the telephone. The Jackson Laboratory at Bar Harbor, Maine, had a similar data bank for mouse information and it was in close contact with a parallel data base outside London under the direction of Dr. Mary Lyon.

The term "informatics" had been coined by Dr. Donald Lindberg, Director of the National Library of Medicine, and it was clear that informatics would play a central role in the genome effort with the accession, organizing, banking, networking, and distribution of the immense amount of information accumulating as the genome effort progressed. Clusters of small data bases have evolved as well as an exponential growth of those just mentioned. The principal resources, however, continue to be the Los Alamos data pool of DNA sequences and the Gene Data Base (GDB) for mapping information (formerly Ruddle's Human Gene Mapping Library, HGML). It is now located at the Johns Hopkins Medical School.

National Coordination, Particularly in the Social Issues

One of the initial and most important roles of the National Center for Human Genome Research, led by Nobel laureate James Watson, and its corresponding directorate in the Department of Energy, led by Dr. David Galas, was to develop guidelines for data accession and storage as well as intercommunication between data bases. For this purpose, the Joint Informatics Task Force was set up under Dr. Dieter Soll of Yale and Dr. Mark Pearson of the DuPont company. They have assembled a detailed report containing chapters dealing with computer technology, analysis of genomic data, collection and management of laboratory data, communication between data bases, and training and development in the informatics of genomics.

Another joint committee, the ELSI, titled a "Working Group" and chaired by Dr. Nancy Wexler, was formed. She was a logical person to head this group, which is so central to the humane aspects of the overall effort, being a well-trained social scientist from a family with a history of Huntington disease. This genetic abnormality affects brain function in

Table 1.1 Genome Expenditures ($ millons/annum)

	Fiscal Year							
	'87	**'88**	**'89**	**'90**	**'91**	**'92**	**'93**	**'94**
National Institutes of Health	0	17	28	60	88	105	106	129
Department of Energy	5	12	18	28	47	61	65	?
Department of Agriculture	0	0	0	0	15	15	15	?

midlife and is inherited in a dominant fashion (see Appendix, Figures 4 and 5), meaning one-half of the children are at risk. Thus the molecular diagnosis of those at risk versus those not at risk, which can now be done with much certainty, poses a major problem in medical and emotional management, including possible amniocentesis and the pros and cons of abortion if the child is diagnosed as a probable carrier of the disease gene (Chapter 9). Dr. Wexler made major contributions to the efforts to make a molecular diagnosis by assembling an extremely large kindred in Venezuela, which was used for the linkage analysis to map the disease gene.

The ELSI group considers its major priority issues relating to the validity of and access to genetic tests and related information, fairness in the use of the information in employment and insurance, the storage, confidentiality, and accessibility of genetic information, and, lastly, information and education prepared for the public to deal with the ethical, legal, and social problems created by the availability of the genetic information.

Finances

The growth of the American genome initiative is reflected in the federal funds added to the governmental support of basic biomedical science, which approximates several billion dollars annually (see Table 1.1). Of the 1991 NIH funds of $88 million, $28 million was devoted to human mapping and sequencing. Of the remainder, $8 million to the mouse, $3 million to the bacterium *E. coli,* $2 million to the worm *C. elegans,* $1 million to the fruit fly, *Drosophila,* $4 million to all other microorganisms, and $500,000 to plants and yeast. These amounts constitute only the added funds appropriated by Congress for the new effort and do not include the approximate 1 to 2 billion dollars of the total 1991 NIH budget of $8 billion ($10 billion, 1993) earmarked for biomedical science related to genetics. One might even say that the NIH and Energy health science budgets deal entirely with genetics, because genes and their variations (alleles) underlie all life, healthy or diseased.

On an international scale, the Human Gene Organization (HUGO), a body of geneticists and molecular biologists elected in an academy fashion by nomination and subsequent voting by the membership, was formed and incorporated in Switzerland in September 1988 (McKusick 1989b). HUGO's role is to coordinate communication and data exchange across international boundaries. HUGO currently hosts the single chromosome mapping workshops and has assumed responsibility for the biannual international Human Gene Mapping conferences, the last being held in Baltimore in September 1992 with Sir Walter Bodmer, President of HUGO, serving as its chairman. A number of other organizations have arisen, including national groups in the United Kingdom, France, the Commonwealth of Russian Republics (formerly the USSR), Japan, and others. So far, the United States has made the major contributions in funding and information produced, probably well over three-fourths of the total effort. However, in the consideration of the moral and ethical issues, some of the first meetings were held in Europe, particularly (and with good historical reasons) in Germany. Thomas Caskey, then of Houston, a molecular-oriented human geneticist replaced Bodmer as president of HUGO in 1993.

The Plan

The DOE and NIH jointly launched a 5-year plan to start in fiscal 1991. In essence, this was to be the first leg of a more ambitious series of three 5-year plans with the ultimate objective that the entire 3 billion bases of a single copy of the human genome would be mapped and sequenced. The first 5-year proposal, as spelled out in the formal document, was to saturate the genome with markers spaced originally 10 cM apart, meaning about 300 to 400 markers, and, subsequently, at 1 cM, necessitating ten times that number. Thus the entire genome would be mapped at equally spaced and uniquely specific loci. Sequencing, on the other hand, would focus on areas of interest mainly directed to human disease. Currently in GenBank, the Los Alamos gene data bank operated by the DOE, there are stored about 100 million bases of human data, much of which is redundant, and, unfortunately, with many errors in both original analysis and subsequent data transfer. Much effort in the first 5 years will be devoted to more efficient, more accurate, and cheaper technologies for sequencing. There is also, even in the best hands, a 1% error. More complicating are the many minor normal variations (polymorphisms), with one human sequence differing on average every several hundred bases from another.

Thus several genomes with obvious redundancy must be analyzed to sort out laboratory errors from normal variations or from those causing disease. The big problem is to differentiate a benign, insignificant polymorphism from a difference of a single base that could lead, for example, to sickling of red cells or to a viscous, pathological, bronchial mucus secretion (cystic fibrosis).

The second 5 years will be devoted to closing the map with the final markers at a 1 cM level, with, it is hoped, sequencing being done cheaply, efficiently, and with minimum error, and separating the frequent and widely scattered polymorphisms and other differences between individuals. Many of these polymorphisims will have much anthropological value in helping to delineate human evolutionary development. The last 5 years will be directed to sequencing the remainder of the genome and, again, looking for the many interindividual differences. The entire project is estimated to cost $3 billion (1991 dollars), minuscule in relation to agriculture or defense dollars, yet a large sum of money. It is reasonable to guess that the entire bill may be ten times this amount. Francis Collins, Director of the National Center for Human Genome Research, and David Galas, formerly Associate Director of the Office of Health and Environmental Research of the DOE, revised and updated the plan (Collins and Galas 1993). The results of the first 5 years have far surpassed all expectations, and Collins and Galas conclude that the future has even more to offer to scientific knowledge and the well-being of humanity.

Recent Happenings

It is well beyond the scope of this chapter to try to summarize the developments in the genome project in 1991 and 1992. A dramatic example, though small in scope, is one offshoot, human gene therapy (see Chapter 8). There are, as of January 1993, about three dozen projects worldwide, with one half having immediate life-saving results. Two school girls with a fatal immune disorder (adenosine deaminase deficiency) are back in class with the missing gene having been added to their white cells by researchers at the National Institutes of Health led by French Anderson and Michael Blaese. The location of this gene, its isolation, and its characterization are all results of the genome technology evolving from the genome project.

Two human chromosomes, the male-determining Y and the second smallest autosome 21 have been almost totally mapped. The Y has 60 million bases, and David Page and colleagues at the Whitehead Institute in Boston have characterized and labeled 196 overlapping human Y fragments grown in yeast. Likewise, Daniel Cohen of Paris and a number of international collaborators have identified 191 markers along the 42 million bases of autosome 21, the chromosome that bears Down syndrome, as well as some potential Alzheimer's and other genes of much medical interest.

Thus the yeast artificial chromosome (YAC) technology has been very catalytic to mapping progress, but another technical advance has provided even greater efficiency. As stated early in this chapter, the original variations detected by the RFLP technology were infrequent and not randomly scattered over the genome. The technology itself, as developed by Ray White, Helen Donis-Keller, and others, was slow, expensive, and not as informative as had been hoped. It had, however, shown that mapping was feasible and it brought the project a long way, allowing diseases such as Huntington, cystic fibrosis, Duchenne muscular dystrophy, and many others to be mapped and thus permitting diagnosis and other medical applications. Human beings have scattered through their entire genome, as probably do many other animals, clusters of repeated sets of two, three, or four bases called microsatellites. Alec Jeffries in England had originally used similar clusters (variable number tandem repeats, VNTRs) as markers because they differ so much from location to location in the genome of a given individual and, more importantly, any single microsatellite differs markedly from person to person unless they are related. This technology is another breakthrough—namely, the capacity to use nature's way of amplifying a minuscule sample of DNA (say, from even a single cell) into enough DNA that it can be characterized by sequencing or by probes able to identify the sequence. This process is termed the *polymerase chain reaction,* or PCR. Thus sequence-tagged sites (STSs) can be located up and down the chromosomes with great specificity and efficiency, and most of these sites will be the aforementioned microsatellites. Confusingly, these sites are called by some "simple sequence repeats," or SSRs.

By 1993, there were several major genome "factories," Genethon in Paris directed by Daniel Cohen and two in the United States, one in Cambridge, Massachusetts, led by Eric Lander and the other a consortium based at the University of Iowa. The aim of the Lander group, for example, was to characterize 10,000 STSs in the human genome and from 4,000 to 6,000 in the mouse. The human map would thus have a density of one STS each 300,000 bases, or, in genetic terms, about every 0.3 cM. An immediate fall-out will be the use of these markers to dramatically decrease the potential error in genetic diagnosis, as in amniocentesis.

In October 1992, a consortium of NIH and CEPH in Paris published 1,416 loci covering 95% of the human X chromosome and 92% of the remainder. So the science of the human genome project was well ahead of the proposed time table originally outlined by James Watson and the founding governmental and advisory committees.

The National Institutes of Health, through the Center for Human Genome Research, initiated eleven genome centers between October 1990 and October 1992 in the United States. These included the two "factories" in Cambridge, Massachusetts, and Iowa mentioned earlier. Similarly, the DOE created three centers of its own.

Progress in the entire genome effort has far exceeded expectations and predictions. Some genetic diseases such as myotonic dystrophy have been found to be related to multiple repeats of simple sequences such as CGG, and these repeated sequences are not only variable between individuals but may even be unstable from one generation to another. They have been termed "premutational" and, in some diseases, such as Huntington, this phenomenon explains why disease may become more severe from one generation to another. It should be added that the high variability of these repeats from one person to another has provided an excellent tool for linkage analysis as well as for forensic purposes.

By early 1995, the Los Alamos–located gene base, GenBank, contained more than 90,000 characterized bits of sequence from expressed genes in human beings, about 30,000 from plants, and some 12,000 from invertebrates such as *Drosophila*. Because some genes are expressed in all tissues and others are limited to only specific tissues, Venter and his colleagues at TIGR have more than 300 libraries of small portions of expressed genes from some three dozen human tissues, some embryonic and some cancerous.

These small portions have been labeled EST for expressed sequence tags because they are derived from sequencing the ends of genes expressed in various organs and tissues. In a "Genome Directory" published by *Nature* in 1995, Venter's group (TIGR) reports on some 55,000 ESTs of which only about 10,000 are available in prior public data bases. They contain about 5 million sequenced base pairs, only 0.15% of the human genome (Maddox 1995). The French group Genethon, led by Daniel Cohen, has used the DNA collected by the Centre d'Etudé Polymorphisme Humaine (CEPH) to prepare a physical map to position ESTs along the human chromosome. To do this, they used yeast artificial chromosome (YAC) technology, which involves cloning large pieces of human DNA into yeast and then using overlapping stretches to piece together, like a large jigsaw puzzle, the contiguous elements. Combination of the Venter data with those derived from Genethon is a major step forward in the total genome effort.

A major contribution to the genome project has been the development of *positional cloning* (Collins 1995). This technical breakthrough permits isolation and characterization of a given gene once its approximate location is known by standard genetic mapping techniques. It took more than 8 years to isolate the Huntington gene, but with positional cloning, by mid-1995, approximately 50 disease-related genes had been characterized. These include genes for breast cancer, colon cancer, retinitis pigmentosa, several genes for diabetes and for Alzheimer's, and even a gene directly related to obesity in experimental animals (Zhang 1994).

Again, the genome movement is accelerating at a totally unexpected pace and the sequencing and characterization of the entire human genome will probably be achieved before the year 2000, if not sooner. The impact of this new knowledge on humankind is unpredictable.

Some Problems

With the preceding history as background, it is easy to see that a major new science has arisen. Many think it is the most significant intellectual discovery in humanity's scientific evolution. But along with the new science are now a number of societal questions at all levels. In addition to the intellectual or scientific interests, economic interests are the dominant and possibly the driving force in the development of the new genetic information, but close behind in shaping this development are ethics and the law. However, the last two serve more to check and balance the scientific and economic driving forces. Already major positions have been taken and the endeavor has already attracted sophisticated and concerned humanists as well as the usual spectrum of informed and uninformed whistle-blowers (Kevles and Hood 1993). On the economic side are the many new proprietary issues and even an entirely new industry made up mainly of small biotechnology companies supported by venture capital (Anderson 1993). A number have already been swallowed up by larger firms and a few have already folded.

Currently a major issue is patenting. Craig Venter, a scientist in the intramural NIH program, with the urging of the Director of NIH, Bernardine Healy, applied for patents on 350 bits of DNA in June 1991 and another 2,735 pieces in February 1992. The first application raised a hue and cry among many scientists, including James Watson, with the feeling that patent protection would impede scientific progress. The British and other foreign scientists similarly objected; however, Sidney Brenner, a leading British molecular and developmental biologist, simultaneously applied for patents on a series of similar DNA fragments. In both, the pieces of DNA were beginning sequences of expressed genes (cDNAs), without knowledge of what the gene was, its role, or where it is located. Is this simply a patenting of a piece of anatomy? Will it facilitate or impede progress in further mapping and sequencing of the human genome? In July 1992, Venter signed a contract with a venture-capital investment group to form a private, but not-for-profit, company to continue isolating and characterizing additional cDNA, the Institute for Genome Research (TIGR). The Merck company has given Washington University in St. Louis a $10 million grant to do the same as the Venter group but to provide the data to the public without the requirement of signing off on its commercial exploitation. By 1995, even religious leaders had been brought into the fray, led by Jeremy Rifkin, the renowned whistle-blower, and question the patenting of genes, gene sequences, animal components, and whole animals such as those with added ("transgenic") or deleted ("knockout") genes (Stone 1995).

Politics again took precedent over administrative logic. James Watson was so bitterly opposed to the patenting of human gene sequences that he voiced his concerns at every opportunity, incurring the wrath of Bernardine Healy, his boss. The result was Watson's requested resignation, bringing many in the scientific community to his defense. He was not replaced until April 1993, when Francis Collins, M.D., Ph.D., was finally appointed. In essence, it was another invasion into the control of science by the body politic. Currently, the issues of patenting and rights of ownership of the chemicals, of the animals, and of the involved knowledge are being debated at all levels: scientific, legal, economic, and, of course, from the ethical and social perspectives.

Similarly, who will control an individual's genetic data as well as the anonymous pooled genome data that are critical for epidemiologic and anthropologic studies? Can these data be used in employment, marriage counseling, insurance applications, forensic identification, filing simply as a mechanism for general identification like a social security number, to name a few? Most importantly, who will decide the policy for the data's use? Should they be elected or appointed politicians, civil servants, lawyers, clergy, ethicists, physicians, economists, the public by referenda? In this book, we hope to provide a clear account of morality, so that, no matter who decides, they will do so within morally acceptable limits. We also hope to clarify some conceptual issues, such as what counts as a genetic malady. This clarification, together with the account of morality, should enable morally acceptable public policies to be developed. We do not think that most of the moral problems that arise from the new genetic information will have unique answers, but we do think that there is a range of morally acceptable answers, and we hope to provide a way of determining what that range is.

As world population expands and resources become rate limiting, as is now happening, the economic issues will no doubt dominate, economics meaning the control of resources. It is hoped that even these major problems will be based on a clear, comprehensive, and commonly accepted moral foundation. Again, we do not claim to be able to provide the answers, only to provide a guide for determining which answers are morally acceptable and which are not. Clarifying the conceptual and moral issues involved is thus the purpose of this book.

Morality, Man's Best Friend the Dog, and Some Even Greater Problems Ahead?

One problem has already arisen with diabetes, not the common type afflicting the middle-aged and frequently managed by diet, weight loss, and exercise alone, but the type that afflicts mainly children at the time of adolescence. The science is now at the point where children can be spotted ahead of time with some degree of statistical confidence; there are complicated and potentially dangerous therapies, and the entire problem is very complex emotionally and economically and sorely in need of humanitarian guidance. As big a problem as it is, with about 1 in every 300 American children by age 18 taking insulin daily to stay alive, it may well be diminished by some of the future problems the genome project could introduce. Nevertheless, Type I diabetes is an excellent example of today's interplay between health policies, economics, genetics, and the moral issues these raise—and in a young and productive population. However, let us focus on some impending problems on the horizon, and probably not as far off as most expect.

In the first paragraph of this chapter on the history of the genome, the impact of selective breeding on both the body structure and the behavior of the dog was introduced. Some four or five decades ago, the geneticist C. C. Little and the then Director of the Rockefeller Foundation, Alan Gregg, discussed the heredity of behavioral characteristics. This led to a 13-year study by two scientists well versed in canine psychological testing as well as in genetics, John Paul Scott and John L. Fuller (Scott and Fuller 1965).

Five strains of dogs were selected for their relatively small size as well as for their disparate behaviors: the cocker spaniel, the African basenji, the wire-haired fox terrier, the Shetland sheep dog, and the standard beagle. Crosses and backcrosses were made and the offspring studied for a large number of dog behaviors. Many traits clearly and significantly segregated genetically. For example, the basenji is nippy and is handled on a leash with difficulty; the cocker the reverse. This jitteriness was easily separated in hybrids, and the trait appeared to be inherited by a single gene. Any dog breeder would acknowledge that, even within a given breed, behavior is not only stereotypical for that breed but even variants within the breed are clearly inherited. An untrained Shetland or a border collie pup will "herd" a single child who has left a group of children back to the group and, if the child is reluctant, will crouch and stare the child into submission, just as it would do to a recalcitrant lamb. Likewise a Labrador pup offered a bowl of water will not drink; it will climb in!

With the conservation of genes across evolutionary boundaries, one may ask what are the genes in human beings homologous to those controlling many of the behavioral traits in the dog. Among many other questions, in what centers in the brain are these genes expressed, what is the nature of the proteins, what neurotransmitters or receptors are involved, what is the relative level of their expression? The dog has thirty-nine pairs of chromosomes and some preliminary assignment of genes have been made to simple linkage groups. Two investigators, Jasper Rine at the DOE Lawrence Laboratories in Berkeley, California, and George Brewer at the University of Michigan, have independently started using the most recent technology (PCR and microsatellites) for mapping the dog genome in preparation for a number of studies on dog diseases as well as on their behavioral traits.

These dog studies will permit not only identity of genes in the dog, but also the homologous genes in human beings. On the beneficial side, the search for psychotropic pharmacologic agents will receive a tremendous boost, and it is a reasonable guess that the pharmaceutical industry will capitalize greatly on these studies. On the other hand, there are already enough problems dealing with the moral and ethical issues raised by the genome in the area of somatic-organic diseases such as diabetes, Huntington disease, and cancer. As one enters into the field of the hereditary aspects of human behavior and other brain functions such as intelligence, capabilities to perform unique tasks involving mathematics, music, spatial recognition, or simply physical coordination, the ethical problems are compounded manifold. Consider, for example, the preliminary studies that suggest that inherited factors play a role in determining sexual orientation (Hamer et al. 1993). Gay and lesbian rights activists are concerned that this genetic difference may be construed to mean something that should or could be reversed. Likewise, a significant pedigree has been published reporting a pattern of sex-linked inheritance of pathological aggression in males in the kindred (Anon. 1993).

Should a moratorium be proscribed on all behavioral research and genetics? Some may feel strongly that it should be. On the other hand, when one deals with a patient in and out of severe depression and sees how impotent modern drug therapy may be in certain cases, one is compelled to encourage the science along in the hope that some intervention may specifically alter this pathologic behavior before suicide terminates the problem. The dog genomic studies and even those looking at some very simple behavior patterns in fruit flies will certainly facilitate this area of neuroscience. But moral guidelines are imperative; no rational person will question this obvious conclusion. Many of the following chapters present current problems and the moral issues they raise. The dog behavioral studies just mentioned are an example of the even greater and more complicated problems of the future. To use the vernacular, we are already in deep water with the ethical problems the genome provides relative to physical disease, but far deeper waters lie ahead when we consider intellectual and behavioral inheritance and the genes involved.

References

Anderson, C. Genome project goes commercial. *Science* 259:300–302, 1993.

Anon. Evidence found for a possible "aggression gene." *Science* 260:1722–1723, 1993.

Botstein, D., White, R.L., Skolnick, M., Davis, R.W. Construction of a genetic linkage map in man using restriction fragment polymorphisms. *Am J. Human Genet.* 32: 314–331,1980.

Collins, F., Galas, D. A new five-year plan for the U.S. human genome project. *Science* 262:43–49, 1993.

Collins, F. Positional cloning moves from perditional to traditional. *Nature Genetics* 9:347–350, 1995.

Cook-Deegan, R.M.: The human genome project: formation of federal policies in the United States, 1986–1990. in *Biomedical Politics*. Hanna, K.E. (ed.) National Academy Press. pp. 99–168. 1991.

Cook-Deegan, R.M. *The Gene Wars: Science, Politics and the Human Genome.* New York: W. W. Norton, 1994.

Fleischmann, R.D., Adams, M.D., White, O., et al. Whole genome random sequencing and assembly of *Haemophilus influenzae* Rd. *Science* 269:496–512, 1995.

Genome Directory. *Nature.* In press.

Genome Issue. *Science* 270:349–548, 1995.

Gusella, J.F., Wexler, N.S., Conneally, P.M., et al. A polymorphic DNA marker genetically linked to Huntington's Disease. *Nature* 306:234–238, 1983.

Hamer D.H., Hu, S., Magnuson, V.L., Hu, N., Pattatucci, A.M.L. A linkage between DNA markers on the X chromosome and male sexual orientation. *Science* 261:321–327, 1993.

Hunkapiller, T., Kaiser, R.J., Koop, B.F., Hood, L. Large-scale and automated DNA sequence determination. *Science* 254:59–67, 1991.

Kan, Y.W., Dozy, A.M. Polymorphism of DNA sequence adjacent to human beta-globin structural gene: relationship to sickle mutation." *Proc. Nat. Acad. Sci.* 75:5631–5635, 1978.

Kevles, D.J. *In the Name of Eugenics: Genetics and the Uses of Human Heredity.* Berkeley: University of California Press, 1985.

Kevles, D.J., Hood, L. (eds.) *The Code of Codes: Scientific and Social Issues in the Human Genome Project*. Cambridge: Harvard University Press, 1993, 397 pp.

Koenig, M., Hoffman, E.P., Bertelson, C.J., et al. Complete cloning of the Duchenne muscular dystrophy (DMD) cDNA and preliminary genomic organization of the DMD gene in normal and affected individuals. *Cell* 50:509–517, 1987.

Lee, T. *The Human Genome Project: Cracking the Genetic Code of Life*. New York: Plenum, 1991, 360 pp.

Lee, T.F. *Gene Future: The Promise and Perils of the New Biology*. New York: Plenum, 1993 (in press).

Lindee, M.S. *Suffering Made Real: American Science and the Survivors of Hiroshima*. Chicago: University of Chicago Press, 1995.

Maddox, J. Directory to the human genome. *Nature* 376:459–460, 1995

McKusick, V.A. *Mendelian Inheritance in Man*. Baltimore: Johns Hopkins University Press. 1994.

McKusick, V.A. Mapping and sequencing the human genome. *New Engl. J. Med.* 230:910–915, 1989a.

McKusick, V.A. The human genome organization: history, purposes, and membership. *Genomics* 5:385–387, 1989b.

Muller-Hill, B. The shadow of genetic injustice. *Nature* 362:491–492, 1993.

Reeders, S. T., Breuning, M.H., Davies, K.E., et al. A highly polymorphic DNA marker linked to adult polycustic kidney disease on chromosome 16. *Nature* 317:542–544, 1985.

Riordan, J.R., Rommens, J.M., Kereme, B., et al. Identification of the cystic fibrosis gene: cloning and characterization of complementary DNA. *Science* 245:1066–1073, 1989.

Scott, J., Fuller, J.L. *Dog Behavior: The Genetic Basis*. Chicago: University of Chicago Press, 1965, 468 pp.

Stone, R. Religious leaders oppose patenting genes and animals. *Science* 268:1126, 1995.

Tauber, A.I., Sarkar, S. The human genome project: has blind reductionism gone too far? *Perspectives Biol. Med.* 35:221–236, 1992.

Watson, J.D., Crick, F.H.C. Genetic implications of the structure of deoxynucleic acid. *Nature* 171:737–738, 1953.

Zhang, Y., Proenca, R., Maffel, M., et al. Positional cloning of the mouse *obese* gene and its human homologue. *Nature* 372:425–432, 1994.

General Medical Resources on Genetics

Genbank, Los Alamos National Laboratory, DOE, Los Alamos NM (a computerized repository of DNA sequences derived from many organisms including human beings).

Human Genome Data Base. John Hopkins University, 1830 East Monument St. Baltimore, MD 21205 (a computerized data base).

McKusick, V.A. *Mendelian Inheritance in Man:* Catalogs of autosomal dominant, autosomal recessive, and X-linked phenotypes. 2 volumes. 11th ed. Baltimore: Johns Hopkins University Press, 1994, 3,009 pp. (a comprehensive listing of all known hereditary human diseases and nondisease traits).

O'Brien, S.J. (ed) *Genetic Maps: Locus Maps of Complex Genomes*, 5th ed. Long Island, NY: Cold Spring Laboratory Press, 1990 (a catalogue of chromosomes and gene locations in all organisms so far studied including human beings and subhuman primates).

Scriver, C.R., Beaudet, A.L., Sly, W.S., Valle, D. (eds). *The Metabolic Basis of Inherited Disease,* 3 volumes. 7th ed. New York: McGraw-Hill, 1995, 5,248 pp. (a detailed series of chapters on hereditary diseases in human beings for which the molecular and biochemical bases have been defined).

Thompson, M.W., McInnes, R.R., Willard, H.F. *Genetics in Medicine,* Philadelphia: Saunders, 1991, 500 pp. (a general text for human genetics).

Significant Organizational Milestones in the Evolution of the Genome Project

Office of Technology Assessment. *Mapping Our Genes—the Human Genome Projects: How Big, How Fast?* OTA-BA-373. U.S. Govt. Printing Office, April, 1988.

National Academy of Science: *Report of the Committee on Mapping and Sequencing the Human Genome.* Washington, D.C.: National Academy Press, 1988, 116 pp.

Department of Energy: Health and Environmental Research Advisory Committee (HERAC). *The Human Genome Initiative,* Dept. of Energy, April 1987.

Department of Energy: *The Human Genome,* DOE/ER-0544P, 1992.

National Academy of Science: *DNA Technology in Forensic Science.* Washington, D.C.: National Academy Press, 1993, 200 pp.

Periodicals

Human Genome News, published by NIH and DOE. HGMIS, Box 2008, Oak Ridge National Laboratory, Oak Ridge, TN 37831-6050.

Progress: Human Genome Project, published by the National Center for Human Genome Research, NIH, Bethesda, MD.

Genomics, Academic Press, Orlando, Florida, monthly.

Nature: Genetics, monthly.

Other Books on the Genome

Bishop, J.E., Waldholz, M. *Genome.* New York: Simon & Schuster, 1990, 352 pp.

Jones, S. *The Language of the Genes: Biology, History, and the Evolutionary Future.* New York: Doubleday, 1993.

Wills, C. *Exons, Introns, and Talking Genes: The Science behind the Human Genome Project.* New York: Basic Books, HarperCollins, 1991, 368 pp.

Wingerson, L. *Mapping Our Genes.* New York: Dutton, 1990.

2

Moral Theory and the Human Genome Project

A clear account of our common morality shows that there is far more agreement on moral matters than is usually assumed. Everyone agrees that killing, causing pain, disabling, depriving of freedom, depriving of pleasure, deceiving, breaking promises, cheating, breaking the law, and neglecting one's duties are not morally allowed unless one has an adequate justification. In presenting detailed analyses of rationality, impartiality, and the moral system, we hope to provide a method for determining what counts as an adequate justification and hence to set limits to legitimate moral disagreement.

Introduction

Any useful attempt to resolve the moral problems that may arise because of the new information that is generated by the Human Genome Project requires an explicit, clear, and comprehensive account of morality. Because some of the problems that will be generated by the Human Genome Project seem to be so different from the kinds of moral problems we normally confront, it is likely that many people will find it difficult to apply their intuitive understanding of morality to these problems. This chapter is an attempt to provide a clear and explicit description of our common morality; it is not an attempt to revise it. Common morality does not provide a unique solution to every moral problem, but it always provides a way of distinguishing between morally acceptable answers and morally unacceptable answers; that is, it places significant limits on legitimate moral disagreement.

One reason for the widely held belief that there is no common morality is that the amount of disagreement in moral judgments is vastly exaggerated. Most people, including most moral philosophers, tend to be interested more in what is unusual than in what is ordinary. It is routine to start with a very prominent example of unresolvable moral disagreement—for example, abortion—and then treat it as if it were typical of the kinds of issues on which one must make moral judgments. It may, in fact, be typical of the kinds of issues on which one makes moral judgments, but this says more about the word "issues" than it does about the phrase "moral judgments." Generally, the word "issues" is used when talking about controversial matters. More particularly, the phrase "moral issues" is always used to refer to matters of great controversy. Moral judgments, however, are not usually made on moral issues: we condemn murderers and praise heroic rescuers; we reprimand our children or our neighbor's children for taking away the toys of smaller children; we condemn cheating and praise giving to those in need. None of these are "moral issues," yet they constitute the subject matter of the vast majority of our moral judgments. These moral judgments, usually neglected by both philosophers and others, show how extensive our moral agreement is.

Areas of Moral Agreement

There is general agreement that such actions as killing, causing pain or disability, and depriving of freedom or pleasure are immoral unless one has an adequate justification. Similarly, there is general agreement that deceiving, breaking a promise, cheating, breaking the law, and neglecting one's duties also need justification in order not to be immoral. There are no real doubts about this. There is some disagreement about what counts as an adequate moral justification for any particular act of killing or deceiving, but there is overwhelming agreement on some features of an adequate justification. There is general agreement that what counts as an adequate justification for one person must be an adequate justification for anyone else in the same situation—that is, when all of the morally relevant features of the two situations are the same. This is part of what is meant by saying that morality requires impartiality.

There is also general agreement that everyone knows what kinds of behavior morality prohibits, requires, encourages, and allows. Although it is difficult even for philosophers to provide an explicit, clear, and comprehensive account of morality, most cases are clear enough that almost everyone knows whether or not some particular piece of behavior is morally acceptable. No one engages in a moral discussion of questions such as, "Is it morally acceptable to deceive patients in order to get them to participate in an experimental treatment that one wants to test?" because everyone knows that such deception is not justified. The prevalence of hypocrisy shows that people do not always behave in the way that morality requires or encourages, but it also shows that everyone knows what kind of behavior morality does require and encourage. This is part of what is meant by saying that morality is a public system.

Finally, there is general agreement that the world would be a better place if everyone acted morally and that it gets worse as more people act immorally more often. This explains why it makes sense to try to teach everyone to act morally even though we know that this effort will not be completely successful. Although in particular cases a person might benefit personally from acting immorally (e.g., providing false information in order to get government Medicare payments when there is almost no chance of being found out), even in these cases it would not be irrational to act morally—namely, not to provide this kind of information even though it means one will not get those payments. We know that the providing of such false information is one of the causes of the problems with the health care system, which results in many people suffering. Morality is the kind of public system that every rational person can support. This is part of what is meant by saying that morality is rational.

A Moral Theory

A moral theory is an attempt to make explicit, explain, and, if possible, justify morality—that is, the moral system that people use in making their moral judgments and in deciding how to act when confronting moral problems. It attempts to provide a usable account of our common morality; an account of the moral system that can actually be used by people when they are confronted with new or difficult moral decisions.[1] It must include an accurate account of the concepts of rationality, impartiality, and a public system, not only because they are necessary for providing a justification of morality, but also because they are essential to providing an adequate account of it. Indeed, a moral theory can be thought of as an analysis of the concepts of rationality, impartiality, a public system, and morality itself, showing how these concepts are related to each other. In this book, we hope to use the clear account of morality, or the moral system presented by the moral theory, to clarify and resolve some of the moral problems that have arisen and will arise from the new information that has been and will be gained from the Human Genome Project.

Rationality is the fundamental normative concept. A person seeking to convince people to act in a certain way must try to show that this way of acting is rational—that is, either rationally required or rationally allowed. We use the term "irrational" in such a way that everyone would admit that if a certain way of acting has been shown to be irrational—that is, not even rationally allowed—no one ought to act in that way.[2] But that a way of acting is rationally allowed does not mean that everyone agrees that one ought to act in that way. On the contrary, given that it is often not irrational (i.e., rationally allowed) to act immorally, it is clear that many hold that one should not act in some ways that are rationally allowed. However, there is universal agreement that any action that is not rationally allowed ought not be done; that is, no one ever ought to act irrationally. If rationality is to have this kind of force, the account of rationality must make it clear why everyone immediately agrees that no one ever ought to act irrationally.

To say that everyone agrees that they ought never act irrationally is not to say that people never do act irrationally. People sometimes act without considering the harmful consequences of their actions on themselves; and, although they do not generally do so, strong emotions sometimes lead people to act irrationally. But regardless of how they actually act, people acknowledge that they should not act irrationally. A moral theory must provide an account of rationality such that, even though people do sometimes act irrationally, no one thinks that he ought to act irrationally. It must also relate this account of rationality to morality.

Impartiality is universally recognized as an essential feature of morality. A moral theory must make clear why morality requires impartiality only when one acts in a kind of way that harms people or increases their probability of suffering harm and does not require impartiality when deciding which people to help—for example, which charity to give to. Most philosophical accounts of morality are correctly regarded as having so little practical value because of their failure to consider the limits on the moral requirement of impartiality. That an adequate account of impartiality requires relating impartiality to some group (e.g., as a father is impartial with regard to his children) explains why abortion and the treatment of animals are such difficult problems. People may differ concerning the size of the group with regard to which morality requires impartiality: some holding that this group is limited to actual moral agents; some holding that it should include potential moral agents—for example, fetuses; and still others claiming that it includes all sentient beings—for example, most mammals. We do not think there are conclusive arguments for any of these views. (See Chapter 9.)

Most moral theories, unfortunately, present an oversimplified account of morality. Philosophers seem to value simplicity more than adequacy as a feature of their theories. Partly, this is because they do not usually think that their theories have any practical use. Many are more likely to accept theories that lead to obviously counterintuitive moral judgments than to make their theories complex enough to account for many of our actual considered moral judgments. This has led many in applied ethics to claim to be anti moral theory. They quite rightly regard these very simple kinds of theories as worse than useless. Unfortunately, they seem to accept the false claim of the theorists that all ethical theories must be very simple. Thus they become anti theory and are forced into accepting the incorrect view that moral reasoning is ad hoc or completely relative to the situation.

The correct Aristotelian middle ground is that moral reasoning is not ad hoc; nor is there any simple account of morality that is adequate to account for our considered moral judgments. Any adequate moral theory must recognize that neither consequences nor moral rules, nor any combination of the two, are the only matters that are relevant when one is deciding how to act in a morally acceptable way or in making moral judgments. Other morally relevant features—for example, the relation between the parties involved—were almost universally ignored until feminist ethical theory emphasized them. When these other features change, they change the kind of action involved and thus may change the moral acceptability of the action under consideration even though the consequences and the moral rules remain the same.

Another reason for the current low esteem in which philosophical accounts of morality are held is that most of these accounts present morality as if it were primarily a personal matter. It is as if each person decides for herself not only whether or not she will act morally, but also what counts as acting morally. But everyone agrees that the moral system must be known to everyone who is judged by it, and moral judgments are made on almost all adults. This means that morality must be a public system, one that is known to all responsible adults; all of these people must know what morality requires of them. In order to justify morality, a moral theory must show that morality is the kind of public system that all impartial rational persons support.

Rationality as Avoiding Harms

Rationality is very intimately related to harms and benefits. Everyone agrees that, unless one has an adequate reason for doing so, it would be irrational to avoid any benefit or not to avoid any harm. The present account of rationality, although it accurately describes the way in which the concept of rationality is ordinarily used, differs radically from the accounts normally provided by philosophers in two important ways. First, it starts with irrationality rather than rationality, and second, it defines irrationality by means of a list rather than a formula. The basic definition is as follows: *A person acts irrationally when (s)he acts in a way that (s)he knows (justifiably believes), or should know, will significantly increase the probability that (s)he, or those (s)he cares for, will suffer death, pain, disability, loss of freedom, or loss of pleasure; and (s)he does not have an adequate reason for so acting.*

The close relation between irrationality and harm is made explicit by this definition, because this list also defines what counts as a harm or an evil. Everything that anyone counts as a harm or an evil—for example, thwarted desires, diseases or maladies, and punishment—is related to at least one of the items on this list. All of these items are broad categories; so nothing is ruled out as a harm or evil that is normally regarded as a harm. That everyone agrees on what the harms are does not mean that they all agree on the ranking of these harms. Further, pain and disability have degrees, and death occurs at very different ages; so there is no universal agreement that one of these harms is always worse than the others. Some people rank dying several months earlier as worse than a specified amount of pain and suffering, wheras other people rank that same amount of pain and suffering as worse. Thus, for most terminally ill patients, it is rationally allowed either to refuse death-delaying treatments or to consent to them.

Most actual moral disagreements—for example, whether or not to discontinue treatment of an incompetent patient—are based on a disagreement on the facts of the case, such as how painful the treatment would be and how long it would relieve the painful symptoms of the patient's disease. Differences in the rankings of the harms accounts for most of the rest—for example, how much pain and suffering is it worth to cure some disability? Often the factual disagreements about prognoses are so closely combined with different rankings of the harms involved that they cannot be distinguished. Further complicating the matter, the probability of suffering any of the harms can vary from insignificant to almost certain, and people can differ in the way that they rank a given probability of one harm against different probabilities of different harms. Disagreement about involuntary commitment of people with mental disorders that make them dangerous to themselves involves a disagreement about both what percentage of these people would die if not committed and whether a significant probability (say, 10%) of death within one week compensates for a 100% probability of three to five days of a very serious loss of freedom and a significant probability (say, 30%) of long-term mental suffering. Actual cases usually involve much more uncertainty about outcomes as well as the rankings of many more harms. Thus complete agreement on what counts as a harm or evil is compatible with considerable disagreement on what counts as the lesser evil or greater harm in any particular case.

If a person knowingly makes a decision that involves an increase in the probability of herself suffering some harm, her decision will be irrational unless she has an adequate reason for that decision. Thus, not only what counts as a reason, but also what makes a reason adequate must be clarified. *A reason is a conscious belief that one's action will help anyone, not merely oneself or those one cares about, avoid one of the harms, or gain some good—namely, ability, freedom, or pleasure—and this belief is not seen to be inconsistent with one's other beliefs by almost everyone with similar knowledge and intelligence.* What was said about evils or harms in the preceding paragraph also holds for the goods or benefits mentioned in this definition of a reason. Everything that people count as a benefit or a good (e.g., health, love, and friends) is related to one or more of the items on this list or to the absence of one or more of the items on the list of harms. Complete agreement on what counts as a good is compatible with considerable disagreement on whether one good is better than another or whether gaining a given good or benefit adequately compensates for suffering a given harm or evil.

A reason is adequate if any significant group of otherwise rational people regard the harm avoided or benefit gained as at least as important as the harm suffered. People are otherwise rational if they do not knowingly suffer any avoidable harm without some reason. No rankings that are held by any significant religious, national, or cultural group count as irrational—for example, the ranking by Jehovah's Witnesses of the harms that would be suffered in an afterlife as worse than dying decades earlier than one would if one accepted a transfusion is not an irrational ranking. Similarly, psychiatrists do not regard any beliefs held by any significant religious, national, or cultural group as delusions or irrational beliefs; for example, the belief by Jehovah's Witnesses that accepting blood transfusions will have bad consequences for one's afterlife is not regarded as an irrational belief or delusion. The intent is to not rule out as an adequate reason any relevant belief that has any plausibility; the goal is to count as irrational actions only those actions on which there is close to universal agreement that they should not be done.

Any action that is not irrational is rational. This results in two categories of rational actions, those that are rationally required and those that are merely rationally allowed. Because no action will be irrational if one has a relevant religious or cultural reason for doing it, and that reason is taken as adequate by a significant group of people, in what follows we shall ignore particular religious or cultural beliefs by assuming that the persons involved have no beliefs that are not commonly held. Given this assumption, an example of a rationally required action—that is, an action that it would be irrational not to do—would be taking a proven and safe antibiotic for a life-threatening infection. On the same assumption, refusing a death-delaying treatment for a painful terminal disease will be a rationally allowed action—that is, an action that it is irrational neither to do nor not to do. These two categories share no common feature except that they are both not irrational. This account of rationality has the desired result that everyone who is regarded as rational always wants himself and his friends to act rationally. Certainly, on this account of rationality, no one would ever want himself or anyone for whom he is concerned to act irrationally.

Although this account of rationality may sound obvious, it is in conflict with the most common account of rationality, where rationality is limited to an instrumental role. A rational action is often defined as one that maximizes the satisfaction of all of one's desires, but without putting any limit on the content of those desires. This results in an irrational action being defined as any action that is inconsistent with such maximization. But unless desires for any of the harms on the list are ruled out, it turns out that people would not always want those for whom they are concerned to act rationally. If a genetic counselor has a young patient who, on finding out that he has the gene for Huntington disease, becomes extremely depressed and desires to kill himself now, more than twenty years before he will become symptomatic, no one will encourage him to satisfy that desire even if doing so will maximize the satisfaction of his present desires. Rather, everyone concerned with him will encourage him to seek counseling. They will all hope that he will be cured of his depression and then come to see that he has no adequate reason to deprive himself of twenty good years of life.[3] That rationality has a definite content and is not limited to a purely instrumental role (for example, acting so as to maximize the satisfaction of all one's desires) conflicts with most accounts of rational actions, both philosophical and nonphilosophical.[4]

Scientists may claim that both of these accounts of rationality are misconceived. They may claim that, on the basic account of rationality, it is not primarily related to actions at all, but rather rationality is reasoning correctly. Scientific rationality consists of using those scientific methods best suited for discovering truth. Although I do not object to this account of rationality, I think that it cannot be taken as the fundamental sense of rationality. The account of rationality as avoiding harms is more basic than that of reasoning correctly, or scientific rationality. Scientific rationality cannot explain why it is irrational not to avoid suffering avoidable harms when no one benefits in any way. The avoiding-harm account of rationality does explain why it is rational to reason correctly and to discover new truth—namely, because doing so helps people to avoid harms and to gain benefits.

Rationality, Morality, and Self-Interest

Although morality and self-interest do not usually conflict, the preceding account of rationality makes clear that, when they do conflict, it is not irrational to act in either way. Although this means that it is never irrational to act contrary to one's own best interests in order to act morally, it also means that it is never irrational to act in one's own best interest even though this is immoral. Further, it may even be rationally allowed to act contrary to both self-interest and morality if, for example, friends, family, or colleagues benefit. This is often not realized, and some physicians and scientists believe that they cannot be acting immorally if they act to benefit others and contrary to their own self-interest. This leads some to immorally cover up the mistakes of their colleagues, believing that they are acting morally, because they, themselves, have nothing to gain and are even putting themselves at risk.

Although some philosophers have tried to show that it is irrational to act immorally, this conflicts with the ordinary understanding of the matter. There is general agreement, for example, that it may be rational for someone to deceive a client about a mistake that one's genetic counseling facility has made, even if this is acting immorally. Neither in this chapter nor anywhere in this book do we attempt to provide the motivation for one to act morally. That motivation primarily comes from one's concern for others, together with a realization that it would be arrogant to think that morality does not apply to oneself and one's colleagues in the same way that it applies to everyone else. Our attempt to provide a useful guide for determining what ways of behaving are morally acceptable presupposes that the readers of this book want to act morally.

Impartiality

Impartiality, like simultaneity, is usually taken to be a simpler concept than it really is. Einstein showed that one cannot simply ask whether A and B occurred simultaneously, one must ask whether A and B occurred simultaneously with regard to some particular observer, C. Similarly, one cannot simply ask if A is impartial, one must ask whether A is impartial with regard to some group in a certain respect. The following analysis of the basic concept of impartiality shows that to fully understand what it means to say that a person is impartial involves knowing both the group with regard to which her impartiality is being judged and the respect in which her actions are supposed to be impartial with regard to that group. *A is impartial in respect R with regard to group G if and only if A's actions in respect R are not influenced at all by which members of G benefit or are harmed by these actions.*

The minimal group toward which morality requires impartiality consists of all moral agents (those who are held morally responsible for their actions), including oneself, and former moral agents who are still persons (incompetent but not permanently unconscious patients). This group is the minimal group because everyone agrees that the moral rules—for example, "Do not kill" and "Do not deceive"—require acting impartially with regard to a group including at least all of these people. Further, in the

United States and the rest of the industrialized world, almost everyone would include in the group toward whom the moral rules require impartiality infants and older children who are not yet moral agents. However, the claim that moral rules require impartiality with regard to any more-inclusive group is more controversial. Many hold that this group should not be any more inclusive, whereas many others hold that this group should include all potential moral agents, whether sentient or not—for example, a fetus from the time of conception. (See Chapter 9 for further discussion of how this issue affects the controversy concerning abortion.) Still others hold that this group should include all sentient beings—that is, all beings who can feel pleasure or pain, whether potential moral agents or not—for example, all mammals.

The debates about abortion and animal rights are best understood as debates about who should be included in the group toward which the moral rules require impartiality. Because fully informed rational persons can disagree about who is included in the group toward which morality requires impartiality, there is no way to resolve the issue philosophically. This is why discussions of abortion and animal rights are so emotionally charged and often involve violence. Morality, however, does set limits to the morally allowable ways of settling unresolvable moral disagreements. These ways cannot involve violence or other unjustified violations of the moral rules but must be settled peacefully. Indeed, one of the proper functions of a democratic government is to settle unresolvable moral disagreements by peaceful means.

The respect in which morality requires impartiality toward the minimal group (or some larger group) is when considering violating a moral rule—say, killing or deceiving. Persons are not required to be impartial in following the moral ideals—for example, relieving pain and suffering. The failure to distinguish between moral rules, which can and should be obeyed impartially with respect to the minimal group, and moral ideals, which cannot be obeyed impartially even with regard to this group, is the cause of much confusion in discussing the relation of impartiality to morality. The kind of impartiality required by the moral rules involves allowing a violation of a moral rule with regard to one member of the group (for example, a stranger) only when such a violation would be allowed with regard to everyone else in the group (e.g., friends or relatives). It also involves allowing a violation of a moral rule by one member of the group (say, oneself) only when everyone else in the group (say, strangers) would be allowed such a violation.

Acting in an impartial manner with regard to the moral rules is analogous to a referee impartially officiating a basketball game, except that the referee is not part of the group toward which he is supposed to be impartial. The referee judges all participants impartially if he makes the same decision regardless of which player or team is benefited or harmed by that decision. All impartial referees need not prefer the same style of basketball; one referee might prefer a game with less bodily contact, hence calling more fouls, whereas another may prefer a more physical game, hence calling fewer fouls. Impartiality allows these differences as long as the referee does not favor any particular team or player over any other. In the same way, moral impartiality allows for differences in the ranking of various harms and benefits as long as one would be willing to make these rankings part of the moral system and one does not favor any particular person in the group, including oneself or a friend, over any others when one decides to violate a moral rule or judges whether a violation is justified.

A Public System

A *public system* is a system that, in normal circumstances, has the following two characteristics. First, all persons to whom it applies—that is, those whose behavior is to be guided and judged by that system—understand it—that is, know what behavior the system prohibits, requires, encourages, and allows. Second, it is not irrational for any of these persons to accept being guided and judged by that system. The clearest example of a public system is a game. A game has an inherent goal and a set of rules that form a system that is understood by all of the players—that is, they all know what kind of behavior is prohibited, required, encouraged, and allowed by the game; and it is not irrational for all players to use the goal and the rules of the game to guide their own behavior and to judge the behavior of other players by them. Although a game is a public system, it applies only to those playing the game. Morality is a public system that applies to all moral agents; all people are subject to morality simply by virtue of being rational persons who are responsible for their actions.

In order for morality to be known by all rational persons, it cannot be based on any beliefs that are not shared by all rational persons. Those beliefs that are held by all rational persons (rationally required beliefs) include general factual beliefs such as: people are mortal, can suffer pain, can be disabled, and can be deprived of freedom or pleasure; also people have limited knowledge—that is, people know some things about the world, but no one knows everything. On the other hand, not all rational people share the same scientific and religious beliefs; so no scientific or religious beliefs can form part of the basis of morality itself, although, of course, such beliefs are often relevant to making particular moral judgments. Parallel to the rationally required general beliefs, only personal beliefs that all rational persons have about themselves (e.g., beliefs that they themselves can be killed and suffer pain and so forth) can be included as part of the foundation for morality. Excluded as part of a foundation for morality are all personal beliefs about one's race, sex, religion, and so forth, because not all rational persons share these same beliefs about themselves.

Although morality itself can be based only on those factual beliefs that are shared by all rational persons, particular moral decisions and judgments obviously depend not only on the moral system, but also on factual beliefs about the situation. Most actual moral disagreements are based on a disagreement on the facts of the case, but particular moral decisions and judgments may also depend on the rankings of the harms and benefits. A decision about whether to withhold a proband's genetic information from him involves a belief about the magnitude of the risk—for example, what the probability is of the information leading him to kill himself—and the ranking of that degree of risk of death against the certain loss of freedom to act on the information that would result from withholding that information. Equally informed impartial rational persons may differ not only in their beliefs about the degree of risk, but also in their rankings of the harms involved, and either of these differences may result in their disagreeing on what morally ought to be done.

Morality

Although morality is a public system that is known by all those who are held responsible for their actions (all moral agents), it is not a simple system. A useful analogy is the grammatical system used by all competent speakers of a language. Almost no competent speaker can explicitly describe this system, yet they all know it in the sense that they use it when speaking and in interpreting the speech of others. If presented with an explicit account of the grammatical system, competent speakers have the final word on its accuracy. They should not accept any description of the grammatical system if it rules out speaking in a way that they regard as acceptable or allows speaking in way that they regard as completely unacceptable.

In a similar fashion, a description of morality or the moral system that conflicts with one's own considered moral judgments normally should not be accepted. However, an explicit account of the systematic character of morality may make apparent some inconsistencies in one's own moral judgments. Moral problems cannot be adequately discussed as if they were isolated problems whose solution did not have implications for all other moral problems. Fortunately, everyone has a sufficient number of moral judgments that they know to be both correct and consistent so that they are able to judge whether a proposed moral theory provides an accurate account of morality. Although few, if any, people consciously hold the moral system described in this chapter, we believe that this moral system is used by most people when they think seriously about how to act when confronting a moral problem themselves or in making moral judgment on others.

Providing an explicit account of morality may reveal that some of one's moral judgments are inconsistent with the vast majority of one's other judgments. Thus one may come to see that what was accepted by oneself as a correct moral judgment is in fact mistaken. Even without challenging the main body of accepted moral judgments, particular moral judgments, even of competent people, may sometimes be shown to be mistaken, especially when long accepted ways of thinking are being challenged. In these situations, one may come to see that one was misled by superficial similarities and differences and so was led into acting or making judgments that are inconsistent with the vast majority of one's other moral judgments. For example, today most doctors in the United States regard the moral judgments that were made by most doctors in the United States in the 1950s about the moral acceptability of withholding information from their patients as inconsistent with the vast majority of their other moral judgments. However, before concluding that some particular moral judgment is mistaken, one must show how this particular judgment is inconsistent with most of one's more basic moral judgments. These basic moral judgments are not personal idiosyncratic judgments but are shared by all who accept any of the variations of our common moral system—for example, that it is wrong to kill and cause pain to others simply because one feels like doing so.

Morality has the inherent goal of lessening the amount of harm suffered by those included in the protected group, either the minimal group or some larger group; it has rules that prohibit some kinds of actions (for example, killing) and require others (for example, keeping promises) and moral ideals that encourage certain kinds of actions (for example, relieving pain). It also contains a procedure for determining when it is justified to violate a moral rule—for example, when a moral rule and a moral ideal conflict. Morality does not provide unique answers to every question; rather it sets the limits to genuine moral disagreement. One of the tasks of a moral theory is to explain why, even when there is complete agreement on the facts, genuine moral disagreement cannot be eliminated, but it must also explain why this disagreement has legitimate limits. It is very important to realize that unresolvable moral disagreement on some important issues (for example, abortion) is compatible with total agreement in the overwhelming number of cases on which moral judgments are made.

One of the proper functions of a democratic government is to choose among the morally acceptable alternatives when faced with an unresolvable moral issue. One important task of this book is to show how to determine those morally acceptable alternatives, in order to make clear the limits of acceptable moral disagreement. Within these limits, it may also be important to show that different rankings of harms and benefits have implications for choosing among alternatives. If one justifies refusing to allow job discrimination on the basis of race or gender because one ranks the loss of the opportunity to work as more significant than the loss of the freedom to choose whom one will employ, impartiality may require one to refuse to allow job discrimination against those suffering disabilities because of a genetic condition.

Moral disagreement not only results from factual disagreement and different rankings of the harms and benefits, but also from disagreement about the scope of morality—that is, who is protected by morality. This disagreement is closely related to the disagreement about who should be included in the group toward which morality requires impartiality. Some maintain that morality is only, or primarily, concerned with the suffering of harm by moral agents, whereas others maintain that the death and pain of those who are not moral agents is as important, or almost so, as the harms suffered by moral agents. Abortion and the treatment of animals are currently among the most controversial topics that result from this unresolvable disagreement concerning the scope of morality. Some interpret the moral rule "Do not kill" as prohibiting killing fetuses and some do not. Some interpret the moral rule "Do not kill" as prohibiting killing animals and some do not. But, even if one regards fetuses and animals as not included in the group impartially protected by morality, this does not mean that one need hold they should receive no protection. There is a wide range of morally acceptable options concerning the amount of protection that should be provided to those who are not included in the group toward which morality requires impartiality.

Disagreement about the scope of morality is only one of the factors that affect the interpretation of the rules. Another factor is disagreement on what counts as breaking the rule—for example, what counts as killing or deceiving, even when it is clear that the person killed or deceived is included in the group impartially protected by morality. People sometimes disagree on when not feeding counts as killing or when not telling counts as deceiving. But, though there is some disagreement in interpretation, most cases are clear and there is complete agreement on the moral rules and ideals to be interpreted. All impartial rational persons agree on the kinds of actions that need justification (e.g., killing and deceiving) and the kinds that are praiseworthy (e.g., relieving pain and suffering). Thus all agree on what moral rules and ideals they would include in a public system that applies to all moral agents. These rules and ideals are part of our common conception of morality, because it is our view that a moral theory must explain and, if possible, justify our common conception of morality; it should not, as most moral theories do, put forward some substitute for it.

With regard to (at least) the minimal group, there are certain kinds of actions that everyone considers to be immoral unless one has an adequate justification for doing them. Among these kinds of actions are killing, causing pain, deceiving, and breaking promises. Anyone who kills people, causes them pain, deceives them, or breaks a promise, and does so without an adequate justification, is universally regarded as acting immorally. Saying that there is a moral rule prohibiting a kind of act is simply another way of saying that a certain kind of act is immoral unless it is justified. Saying that breaking a moral rule is justified in a particular situation—for example, breaking a promise in order to save a life—is another way of saying that a kind of act that would be immoral if not justified is justified in this kind of situation. When no moral rule is being violated, saying that someone is following a moral ideal—for example, relieving pain—is another way of saying that he is doing a kind of action regarded as morally good. Using the terminology of moral rules and moral ideals, and justified and unjustified violations, allows us to formulate a precise account of morality, showing how its various component parts are related. We believe such an account may be helpful to those who must confront the problems raised by the information that has been and will be gained from the new genetics.

A Justified Moral System

A moral system that all impartial rational persons could accept as a public system that applies to all rational persons is a *justified moral system*. Like all justified moral systems, the goal of our common morality is to lessen the amount of harm suffered by those protected by it; it is constrained by the limited knowledge of people and by the need for the system to be understood by everyone to whom it applies. It includes rules prohibiting causing each of the five harms that all rational persons want to avoid and ideals encouraging the prevention of each of these harms.

The Moral Rules Each of the first five rules prohibits directly causing one of the five harms or evils:

- Do not kill (equivalent to causing permanent loss of consciousness);

- Do not cause pain (includes mental suffering—for example, sadness and anxiety);
- Do not disable (includes loss of physical, mental, and volitional abilities);
- Do not deprive of freedom (includes freedom to act and from being acted on);
- Do not deprive of pleasure (includes future as well as present pleasure).

The second five rules include those rules that, when not followed in particular cases, usually cause harm, and general disobedience always results in more harm being suffered:

- Do not deceive (includes more than lying);
- Keep your promise (equivalent to Do not break your promise);
- Do not cheat (primarily violating rules of a voluntary activity);
- Obey the law (equivalent to Do not break the law);
- Do your duty (equivalent to Do not neglect your duty). The term "duty" is being used in its everyday sense to refer to what is required by one's role in society, primarily one's job, not as philosophers customarily use it, which is to say, simply as a synonym for "what one morally ought to do."

The Moral Ideals In contrast with the moral rules, which prohibit doing those kinds of actions that cause people to suffer some harm or increase the risk of their suffering some harm, the moral ideals encourage one to do those kinds of actions that lessen the amount of harm suffered (including providing goods for those who are deprived) or decrease the risk of people suffering harm. As long as one avoids violating a moral rule, following any moral ideal is encouraged. In particular circumstances, it may be worthwhile to talk of specific moral ideals: for example, one can claim that there are five specific moral ideals involved in preventing harm, one for each of the five kinds of harms. Physicians seem primarily devoted to the ideals of preventing death, pain, and disability. Genetic counselors may have as their primary ideal preventing the loss of freedom of their clients. (See Chapter 6 for further discussion.) One can also specify particular moral ideals that involve preventing unjustified violations of each of the moral rules. Insofar as a misunderstanding of morality may lead to unjustified violations of the moral rules, providing a proper understanding of morality may also be following a moral ideal.

Although it is not important to decide how specific to make the moral ideals, it is important to distinguish moral ideals from other ideals. Utilitarian ideals involve promoting goods (for example, abilities and pleasure) for those who are not deprived. Such ideals are followed by those who train athletes or who create delicious new recipes. Religious ideals involve promoting activities, traits of character, and so forth, that are idiosyncratic to a particular religion or group of religions. Personal ideals involve promoting some activities, traits of character, etc., which are idiosyncratic to particular persons (for example, ambition) about which there is not universal agreement. Except in very special circumstances, only moral ideals can justify violating a moral rule with regard to someone without her consent.

It is the possibility of being impartially obeyed all of the time that distinguishes the moral rules from the moral ideals. Impartial rational persons favor people following both the moral rules and the moral ideals, but it is only failure to obey a moral rule that requires an excuse or a justification. This account of moral rules and ideals should not be at all surprising. All that is being claimed is that everyone counts certain kinds of actions as immoral—for example, killing, causing pain, deceiving, and breaking promises—unless one can justify doing that kind of act and that no one doubts that acting to relieve pain and suffering is encouraged by morality. That two moral rules can conflict—for example, doing one's duty may require causing pain—makes it clear that it would be a mistake to conclude that one should always avoid breaking a moral rule. Sometimes breaking one of these rules is so strongly justified that not only is there nothing immoral about breaking it, it would be immoral not to break the rule. A physician who, with the rational informed consent of a competent patient, performs some painful procedure in order to prevent much more serious pain or death breaks the moral rule against causing pain but is not doing anything that is immoral in the slightest. In fact, refusing to do the necessary painful procedure, given the conditions specified, would itself be a violation of one's duty as a doctor and thus would need justification in order not to be immoral. It is clear, therefore, that to say that someone has broken a moral rule is not, by itself, to say that anything wrong has been done; it is only to say that some justification is needed.

What Counts as a Violation of a Moral Rule?

As mentioned earlier, there is often a difference in interpretation about what counts as breaking the rule. Sometimes people will disagree whether to consider an action a justified violation of a moral rule, as just described, or not even a violation of a rule. Not every action that results in someone suffering a harm or an evil counts as breaking one of the first five rules. A scientist who discovers that another scientist's important new discovery is, in fact, false may know that publishing this will result in the second scientist feeling bad. But publishing her findings is not a violation of the rule against causing pain. Almost no one would say that it was, but determining whether or not it was depends upon the practices and conventions of the society. Often these are not clear; for example, if a genetic counselor responds to a couple's question and informs them that their fetus has some serious genetic problem (say, trisomy 18), she may know that this will result in their suffering considerable grief. However, if she has verified the information and told them in the appropriately considerate way, then many would say that she did not break the rule against causing pain and her action requires no justification. Indeed, not responding truthfully to their question would be an unjustified violation of the rule against deception. This interpretation is taking the counselor to be acting like the scientist reporting a mistake by another scientist. Others might take the genetic counselor to be acting like the doctor justifiably breaking the rule against causing pain because she is doing so with the consent of the couple and for their benefit. In either case, it is at least a moral ideal to be as kind and gentle in telling that truth as one can. Indeed, many would claim it is a duty of genetic counselors to minimize the suffering caused by providing information about serious genetic problems.

It is quite clear that lying, making a false statement with the intent to deceive, counts as a violation of the rule prohibiting deception, as does any other action that is intentionally done in order to deceive others. But it is not always clear when withholding information counts as deception. Thus it not always clear that one needs a justification for withholding some information—for example, that the husband of the woman whose fetus is being tested did not father that fetus. In scientific research, what counts as deceptive is determined in large part by the conventions and practices of the field or area of research. If it is a standard scientific practice not to report unsuccessful experiments or to smooth the curves, then doing so is not deceptive, even if some people are deceived. However, a practice that results in a significant number of people being deceived is a deceptive practice even if it is a common practice within the field or area—for example, releasing to the public press a premature and overly optimistic account of some genetic discovery, thereby creating false hope for those suffering from the related genetic malady. Recognition that one's action is deceptive is important, because then one realizes that one needs a justification for it or else one is acting immorally.

Justifying Violations of the Moral Rules

Almost everyone agrees that the moral rules are not absolute, that they have justified exceptions; most agree that even killing is justified in self-defense. Further, there is widespread agreement on several features that all justified exceptions have. The first of these involves impartiality. There is general agreement that all justified violations of the rules are such that, if they are justified for any person, they are justified for every person when all of the morally relevant features are the same. The major, and probably only, value of simple slogans like the Golden Rule, "Do unto others as you would have them do unto you," and Kant's Categorical Imperative, "Act only on that maxim that you could will to be a universal law," is as devices to persuade people to act impartially when they are contemplating violating a moral rule. However, given that these slogans are often misleading, it would be better to consider whether an impartial rational person could publicly allow that kind of violation when trying to decide what to do in difficult cases.

There is almost complete agreement that it has to be rational to favor everyone being allowed to violate the rule in these circumstances. Suppose that someone suffering from a mental disorder both wants to inflict pain on others and wants pain inflicted on himself. He favors allowing any person who wants others to cause him pain to cause pain to others, whether or not they want pain inflicted on them. Whether or not this person is acting in accord with the Golden Rule or the Categorical Imperative, it is not sufficient to justify that kind of violation. No impartial rational person would favor allowing those who want pain caused to them to cause pain to everyone else, whether or not they want pain caused to them. The result of allowing that kind of violation would be an increase in the amount of pain suffered with almost no compensating benefit, which is clearly irrational.

Finally, there is general agreement that a violation is justified only if it is rational to favor that violation even if everyone knows that this kind of violation is allowed—that is, the violation must be publicly allowed. A violation is not justified simply if it would be rational to favor allowing everyone to violate the rule in the same circumstances, but only if almost no one knows that it is allowable to violate the rule in those circumstances. For example, it might be rational for one to favor allowing a physician to deceive a patient about his diagnosis if that patient were likely to be upset by knowing the truth, when almost no one knows that such deception is allowed. But that would not make deception in these circumstances justified. It has to be rational to favor allowing this kind of deception when everyone knows that one is allowed to deceive in these circumstances. One must be prepared to publicly defend this kind of deception, if it were discovered. Only the requirement that the violation be publicly allowed guarantees the kind of impartially required by morality.

Not everyone agrees on which violations satisfy these three conditions, but there is general agreement that no violation is justified unless it satisfies all three of these conditions. Allowing for some disagreement while acknowledging the significant agreement concerning justified violations of the moral rules results in the following formulation of the appropriate moral attitude toward violations of the moral rules: *Everyone is always to obey the rule unless an impartial rational person can advocate that violating it be publicly allowed. Anyone who violates the rule when no impartial rational person can advocate that such a violation be publicly allowed may be punished.* (The "unless" clause only means that, when an impartial rational person can advocate that such a violation be publicly allowed, impartial rational persons may disagree on whether or not one should obey the rule. It does not mean that they agree that one should not obey the rule.)

The Morally Relevant Features

When deciding whether an impartial rational person can advocate that a violation of a moral rule be publicly allowed, the kind of violation must be described using only morally relevant features. Because the morally relevant features are part of the moral system, they must be such that they can be understood by all moral agents. This means that any description of the violation that one offers as appropriate to determine whether or not an impartial rational person could publicly allow it must be such that it can be reformulated in a way that all moral agents could understand it. Limiting the way in which a violation can be described makes it easier for people to discover that their decision or judgment is biased by some consideration that is not morally relevant. All of the morally relevant features that we have discovered so far are answers to the following questions. It is quite likely that other morally relevant features will be discovered, but we think that we have discovered the major features. Of course, in any actual situation, it is the particular facts of the situation that determine the answers to these questions, but all of these particular facts can be redescribed in a way that can be understood by all moral agents:

1. What moral rules are being violated?
2. What harms are being (a) avoided, (b) prevented, (c) caused?
3. What are the relevant desires of the people toward whom the rule is being violated? (This explains why a patient's consent to treatment is so important.)
4. What are the relevant rational beliefs of the people toward whom the rule is being violated? (This explains why it is important to provide the patient with adequate information.)
5. Does one have a relationship with the person(s) toward whom the rule is being violated such that one has a duty to violate moral rules with regard to the person(s)? (This explains why a parent or guardian is allowed to make decisions about treatment that cannot be made by the health care team.)

6. What benefits are being promoted?
7. Is an unjustified or weakly justified violation of a moral rule being prevented?
8. Is an unjustified or weakly justified violation of a moral rule being punished?
9. Are there any alternative actions that would be preferable?[5]
10. Is the violation being done intentionally or only knowingly?[6]
11. Is the situation an emergency that no person is likely to plan to be in?[7]

When considering the harms being avoided (not caused), prevented, or caused, and the benefits being promoted, one must consider not only the kind of benefit or harm involved, one must also consider their seriousness, duration, and probability. If more than one person is affected, one must consider not only how many people will be affected, but also the distribution of the harms and benefits. If two violations are the same in all of their morally relevant features, then they count as the same kind of violation. Anyone who claims to be acting or judging as an impartial rational person who holds that one of these violations be publicly allowed must hold that the other also be publicly allowed. This follows from the account of impartiality. However, this does not mean that two people, both impartial and rational, who agree that two actions count as the same kind of violation must always agree on whether or not to advocate that this kind of violation be publicly allowed, because they may differ in their estimate of the consequences of publicly allowing that kind of violation or they may rank the benefits and harms involved differently.

To act or judge as an impartial rational person, one decides whether or not to advocate that a violation be publicly allowed by estimating what effect this kind of violation, if publicly allowed, would have. If all informed impartial rational persons would estimate that less harm would be suffered if this kind of violation were publicly allowed, then all impartial rational persons would advocate that this kind of violation be publicly allowed and the violation is strongly justified; if all informed impartial rational persons would estimate that more harm would be suffered, then no impartial rational person would advocate that this kind of violation be publicly allowed and the violation is unjustified. However, impartial rational persons, even if equally informed, may disagree in their estimate of whether more or less harm will result from this kind of violation being publicly allowed. When this happens, even if they are impartial, they will disagree on whether or not to advocate that this kind of violation be publicly allowed and the violation counts as weakly justified. Sometimes, primarily in a consideration of the actions of governments, it is also appropriate to consider not only the harms, but also the benefits that would result from this kind of violation being publicly allowed.

Disagreements in the estimates of whether a given kind of violation being publicly allowed will result in more or less harm may stem from two distinct sources. The first is a difference in the rankings of the various kinds of harms. If someone ranks a specified amount of pain and suffering as worse than a specified amount of loss of freedom and someone else ranks them in the opposite way, then, although they agree that a given action is the same kind of violation, they may disagree on whether or not to advocate that this kind of violation be publicly allowed. The second is a difference in estimates of how much harm would result from publicly allowing a given kind of violation, even when there seems to be no difference in the rankings of the different kinds of harms. These differences may stem from differences in beliefs about human nature or about the nature of human societies. Insofar as these differences cannot be settled by any universally agreed upon empirical method, such differences are best regarded as ideological. The disagreement about the acceptability of voluntary active euthanasia of patients with terminal illnesses is an example of such a dispute. People disagree on whether publicly allowing voluntary active euthanasia will result in various bad consequences, including significantly more people dying sooner than they really want to. However, it is quite likely that most ideological differences also involve differences in the rankings of different kinds of harms; for example, does the suffering prevented by voluntary active euthanasia rank higher or lower than the earlier deaths that might be caused? But sometimes there seems to be an unresolvable difference when a careful examination of the issue shows that there is actually a correct answer.

Applying Morality to a Particular Case

A genetic counselor may claim that deception about a diagnosis (say, of Huntington disease in a young adult) to avoid causing a specified degree of anxiety and other mental suffering is justified. He may claim that withholding unpleasant findings in these circumstances will result in less overall harm being suffered than if such deception were not practiced. He may hold that patients are often not able to deal with bad news and are very unlikely to find out about the deception. Thus he may claim that this kind of deception actually results in patients suffering less harm than if they were told the truth. However, another genetic counselor may claim that deception, no matter how difficult it will be for the client to accept the facts or how confident the counselor is that the deception will not be discovered, is not justified. The latter may hold that deception of this kind will actually increase the amount of harm suffered because patients will be deprived of the opportunity to make decisions based upon the facts and that if they do find out about the deception not only will they have less faith in statements made by the counselor, they will also have less faith in statements made by other health care providers, thus increasing the amount of anxiety and suffering. Thus there is a genuine empirical dispute about whether withholding bad news from patients is likely to increase or decrease the amount of harm suffered. Which of these hypotheses about the actual effects of deception in the particular circumstances is correct we do not know, but if one is concerned with the moral justifiability of such deception it does not matter.

The morally decisive question is not "What are the consequences of this particular act?" but rather "What would be the consequences if this kind of deception were publicly allowed?" Neither counselor has taken into account that a justifiable violation against deception must be one that is publicly allowed—that is, one that everyone knows is allowed. Once one realizes that in making a moral decision one must consider the consequences if everyone knows that it is allowable to deceive in certain circumstances (e.g., to withhold bad news in order to avoid anxiety and other mental suffering) then the loss of trust involved will obviously have worse consequences than if everyone knew that such deception was not allowed. It is only by concentrating on the results of one's own deception, without recognizing that morally allowed violations for oneself must be such that everyone knows that they are morally allowed for everyone, that one could be led to think that such deception was justified. Consciously holding that it is morally allowable for oneself to deceive others in this way although, of course, one would not want everyone to know that everyone is morally allowed to deceive others in the same circumstances, is exactly what is meant by arrogance—namely, the arrogating of exceptions to the moral rules for oneself that one would not want everyone to know are allowed for all. This arrogance is clearly incompatible with the kind of impartiality that morality requires with regard to obeying the moral rules.

The most popular guide to conduct now used by those in the field of bioethics, and especially genetic counseling, is what is called "principlism." This approach uses several principles, the most popular being (1) autonomy, (2) beneficence, (3) justice, and (4) nonmaleficence. Sometimes (5) confidentiality or (6) fidelity is added to this list or substituted for one of the others. Because this approach is so popular in bioethics, we shall devote the entire next chapter to a critique of it, from both a theoretical and a practical point of view. We will then examine a detailed application of a principlist approach to some ethical dilemmas arising in genetic counseling with regard to Huntington disease and show how our approach, using the account of the moral system contained in this chapter, provides more useful guidance in dealing with these dilemmas.

Contrasts with Other Systems for Guiding Conduct

For those who are concerned with the philosophical foundations of bioethics, it may clarify our account of the moral system to compare it with the views put forward by many contemporary followers of Immanuel Kant (1724–1804) and John Stuart Mill (1806–1873). The Kantian Categorical Imperative, "Act only on that maxim whereby you can at the same time will that it be a universal law of nature," and Mill's Utilitarian Greatest Happiness Principle, "Act so as bring about the greatest happiness for the greatest number," are two of the most popular and influential moral philosophical slogans. But these slogans, though often cited, are inadequate, by themselves, to provide a useful moral guide to conduct. It is not fair to Kant and Mill or their contemporary followers to compare these slogans with our account of the moral system sketched in this chapter, because Kant and Mill and their contemporary followers have far more to say than simply working out the consequences of these slogans. However, popular use of these slogans, especially in medical contexts, is often as simple as we shall characterize it. Further, neither Kant nor Mill nor their contemporary followers provide a list of morally relevant features; that is, there is little effort devoted to providing plausible accounts of how one determines whether two violations count as violations of the same kind for the purpose of moral evaluation.

On a popular interpretation of a Kantian deontological system, one should never act in any way that one cannot will to be a universal law. If it would be impossible for everyone always to do a specific kind of action, then everyone is prohibited from doing that kind of action. For example, that it is impossible for everyone always to make lying promises (for then there could be no practice of promising) is what makes it morally prohibited to make lying promises. On the moral system, one is prohibited from doing a kind of action only if, given the morally relevant facts, no impartial rational person would publicly allow that kind of action. A Kantian system seems to rule out ever making lying promises, whereas our common morality allows the making of lying promises in some circumstances—for example, when it is necessary to make the lying promise to prevent a harm sufficiently great that less overall harm would be suffered even if everyone knew such lying promises were allowed.

On a popular interpretation of a Utilitarian or consequentialist system (Bentham and Mill), one not only may, but should, violate any rule if the foreseeable consequences of that particular violation, including the effects on future obedience to the rule, are better than the consequences of not violating the rule. A consequentialist system is concerned only with the foreseeable consequences of the particular violation, not with the foreseeable consequences of that kind of violation being publicly allowed. But, on our moral system, it is precisely the foreseeable consequences of that kind of violation being publicly allowed that are decisive in determining whether or not it is morally allowed. The consequences of the particular act are important only in determining the kind of violation under consideration. A consequentialist system favors cheating on an exam if one were certain that one would not get caught and no harm would result from that particular violation of the rule against cheating. Assuming that the exams serve a useful function, our moral system would not allow this kind of violation of the rule against cheating, because if this kind of violation were publicly allowed, it would make it pointless to have exams.

According to consequentialism, the only morally relevant features of an act are its consequences. It is, paradoxically, the kind of moral theory usually held by people who claim that they have no moral theory. Their view is often expressed in phrases such as: "It is all right to do anything as long as no one gets hurt," "It is the actual consequences that count, not some silly rules," or "What is important is that things turn out for the best, not how one goes about making that happen." According to Classical Utilitarianism (Bentham and Mill), the only relevant consequences are pleasure and pain. That act is considered morally best which produces the greatest balance of pleasure over pain. On our moral system, pleasure and pain are not the only consequences that count, and it is not the consequences of the particular violation that are decisive in determining its justifiability, but rather the consequences of publicly allowing such a violation.

The moral system differs from a Kantian system and resembles a consequentialist system in that it has a purpose, and consequences are explicitly taken into consideration. It resembles a Kantian system and differs from a consequentialist system in that morality must be a public system in which rules are essential. The role of impartiality also differs. The Kantian system requires all of one's actions to be impartial, and consequentialist systems require one to regard the interests of everyone impartially. Morality does not require impartiality with regard to all of one's actions, it requires impartiality only with respect to obeying the moral rules. Nor does morality require one to regard the interests of everyone impartially; it only requires that one act impartially when violating a moral rule. Indeed, it is humanly impossible to regard the interests of everyone impartially, when concerned with all those in the minimal group. Impartiality with respect to the moral ideals (Kant would call these imperfect duties) is also humanly impossible. That all the moral rules are or can be taken as prohibitions is what makes it humanly possible for them to be followed impartially. The public nature of morality and the limited knowledge of rational persons help to explain why impartial obedience to the moral rules is required to achieve the point of morality, lessening the suffering of harm. Morality also differs from both systems in that it does not require all moral questions to have unique answers but explicitly allows for a limited area of disagreement among equally informed impartial rational persons.

Endnotes

1. A more extended account of morality, and of the moral theory that justifies it, is contained in *Morality: A New Justification of the Moral Rules*, by Bernard Gert, Oxford University Press, 1988, 317 pp. (paperback), 1989.

2. We are aware that the terms "rational" and irrational" are used in many different ways—for example, "irrational" means spontaneous. However, we think that there is a basic concept of rationality and that is the one that we are attempting to describe.

3. See Irrationality and the DSM-III-R definition of mental disorder, *Analyze & Kritik*, Jahrgang 12, Heft 1, July 1990, pp. 34–46, by Bernard Gert.

4. See Rationality, human nature, and lists, *Ethics*, Vol. 100, No. 2, January 1990, pp. 279–300 and Defending irrationality and lists, *Ethics*, Vol. 103, No. 2, January 1993, pp. 329–336, by Bernard Gert.

5. This involves trying to find out if there are any alternative actions such that either they would not involve a violation of a moral rule or the violations would differ in some morally relevant features, especially, but not limited to, the amount of evil, caused, avoided, or prevented.

6. There are many other questions: for example, Is the violation being done (a) voluntarily or because of a volitional disability? (See Free will as the ability to will, *Nous*, Vol. 13, No. 2, May 1979, pp. 197–217, Bernard Gert and Timothy Duggan. Reprinted in *Moral Responsibility*, edited by John Martin Fisher, 1986.) (b) freely or because of coercion? (c) knowingly or without knowledge of what is being done? (d) is the lack of knowledge excusable or the result of negligence? whose answers will affect the moral judgment that some people will make. The primary reason for not including answers to these questions as morally relevant features is that our goal in listing morally relevant features is to help those who are deciding whether or not to commit a given kind of violation; so we did not want to include those features that are solely of value in judging violations that have already been committed and cannot be used in deciding how to act. For questions a, b, c, and d, one cannot decide whether or not to commit one rather than another of these kinds of violations, hence they are not useful in deciding how to act.

Although one does not usually decide whether or not to commit a violation intentionally or only knowingly, sometimes that is possible. For violations that are alike in all of their other morally relevant features, a person might not publicly allow a violation that was done intentionally but might publicly allow a violation that was not done intentionally, even though it was done knowingly. For example, many people would publicly allow nurses to administer morphine to terminally ill patients in order to relieve pain even though everyone knows it will hasten the death of the patient but, with no other morally relevant changes in the situation, they would not allow nurses to administer morphine in order to hasten the death of the patient. This distinction explains what seems correct in the views of those who endorse the doctrine of double effect. I think that such a distinction may also account for what many regard as a morally significant difference between lying and other forms of deception, especially withholding information. Nonetheless, it is important to remember that many, perhaps most, violations that are morally unacceptable when done intentionally are also morally unacceptable when done only knowingly.

7. We are talking about the kind of emergency situation that is sufficiently rare that no person is likely to plan or prepare for being in it. This is a feature that is necessary to account for the fact that certain kinds of emergency situations seem to change the moral judgments that many would make even when all of the other morally relevant features are the same. For example, in an emergency when a large number of people have been seriously injured, doctors are morally allowed to abandon patients who have a very small chance of survival in order to take care of those with a better chance, in order that more people will survive. However, in the ordinary practice of medicine, they are not morally allowed to abandon patients with poor prognoses in order to treat those with better prognoses, even if doing so will result in more people surviving.

3

Concerning the Inadequacies of Principlism

Principlism has been the dominant theoretical influence in biomedical ethics in the past two decades. It includes chiefly the principles of nonmaleficence, justice, beneficence, and autonomy. This chapter argues that principlism is misleading and inadequate. This is a detailed and explicit argument undertaken for conceptual and practical reasons. The disadvantages of principlism and the advantages of our account of morality are argued here in the abstract; they are concretely illustrated by comparing a principlist discussion of cases in Chapter 4 with our discussion of the same cases in Chapter 5.

Introduction

In the preceding chapter, we presented our own account of morality, which in effect is a systematic description of the basic morality by which we all live. We did not invent a morality; we did not create a new morality. Rather we spelled out as consistently and rigorously as the subject matter allows the underlying foundations and the embedded moral rules and ideals of ordinary morality. This current chapter is a chapter of transition from the moral theory itself to the use of it in practice, in particular in dealing with moral problems in genetic counseling.

The reason a transition would be helpful to the reader is that, even though we present an account of ordinary morality as it is actually practiced, other accounts of moral reasoning have come to prevail in the biomedical sphere. In particular, the dominant "theory" is that which we have labeled "principlism." It has pervaded almost all of biomedical ethics for more than a decade. Because it has been so pervasive, we need to underline our own significant differences from it, not only to argue for our approach being more adequate and useful, but also to minimize the chances that aspects of principlism might unwittingly and automatically be read into our own. This current chapter of conceptual "ground clearing" will allow us to deal with the cases in the next two chapters in a more focused manner. There we will illustrate precisely how our account is clearer, more adequate, and more useful.

In this current chapter, we will follow the time-honored tradition of theory replacement. That is, we will point out how our theory (1) overcomes the inadequacies of the dominant theory, (2) accounts for what is good in the dominant theory, and (3) is more readily usable, understandable, and intuitively correct than the dominant theory.

Principlism is most commonly characterized by citing four principles—autonomy, beneficence, nonmaleficence, and justice, which constitute the core of its account of biomedical ethics. So entrenched is this "theory," that clinical moral problems are often grouped (for conferences, papers, and books) according to which principle is deemed necessary for solving them. It has become fashionable and customary to cite one or another of these principles as the key for resolving a particular biomedical ethical problem. Throughout much of the medical bioethical literature, authors seem to believe that they have brought theory to bear on the problem under consideration insofar as they have simply mentioned one or more of the principles. Thus, not only do the principles presumably lead to the solutions, but they are also treated by many as the ultimate grounds of appeal.

We will be making our case by examining the leading account of principlism—namely, that of Beauchamp and Childress,[1] as manifested in their book *Principles of Biomedical Ethics*, in all its editions. Their account is the very best the position has to offer, and it is their account that has so pervaded the world of biomedical ethics. For many years, it has provided the conceptual framework of the Georgetown Intensive Bioethics Course, a 1-week summer course that has been attended by thousands from the United States as well as from other countries. Beauchamp and Childress's book is outstanding in its insights about medicine and ethics; it is unsurpassed for its sensitivity to important issues and relevant subtleties. Our criticism is focused only on the authors' misleading theoretical description of what they are doing. We are concerned that this misleading account of moral reasoning may sometimes lead to the wrong answers to complex problems. For the sake of background on principlism's pervasive influence, we will look briefly at the "Belmont Report" which seems to be progenitor of the principles.

The Principles in Historical Context

The principles emerged from the work of the National Commission for the Protection of Human Subjects of Biomedical and Behavioral Research, which was created by Congress in 1974. One of the charges to the Commission was to identify the basic ethical principles that should underlie the conduct of biomedical and behavioral research involving human subjects and to develop guidelines that should be followed to assure that such research is conducted in accordance with those principles.[2]

At that time, there was some frustration over the many and various rules for research that were spelled out in the extant codes covering research using human subjects. These codes included the Nuremberg Code of 1947, the Helsinki Declaration of 1964 (revised in 1975), and the 1971 Guidelines issued by the (then) Department of Health, Education, and Welfare. (The *Guidelines* were codified into Federal Regulations in 1974.) The assortment of rules seemed at times inadequate, conflicting, and difficult to apply. It therefore became part of the Commission's charge to formulate "broader ethical principles [to] provide a basis on which specific rules may be formulated, criticized and interpreted."[3]

The higher level of generality was achieved by the Commission and articulated as three ethical principles: the principle of respect for persons, the principle of beneficence, and the principle of justice. These principles constituted the "Belmont Report," so named because their articulation was the culmination of intense discussions that took place at the Smithsonian Institution's Belmont Conference Center. In effect, these principles sought to frame in a more general and useful way the moral concerns that underlay the (occasionally) diverse, uneven, ambiguous, and conflicting rules constituting the various ethical codes related to research on human subjects.

The work of the Commission was significant. It was insightful and helpful; it elegantly captured in a more general way the basic moral concerns haltingly expressed in the miscellaneous codes. The Commission also went on to delineate some of the more practical consequences of the principles. From the principle of respect for persons came attention to autonomy (which for them seems more like what we would call "competence") and to informed consent. From the principle of beneficence came the obligation not to harm and to maximize benefits over risks. From the principle of justice came attention to fairness in the distribution of the benefits and burdens of research.

These principles were clearly intended to be generalized guides for protecting human beings as subjects in biomedical and behavioral research. Also, they seem less to have been derived from a theory of any sort and more to have been either abstractions from moral rules or expressions of particular ethical concerns. In a summary fashion, the principles generalize and encapsulate the more immediate moral considerations especially applicable to research using human subjects. Very likely these formulations additionally accomplished a crucial maneuver for the Commission.

They made possible a consensus in a setting where a moral theory would probably never have been agreed upon.

From these beginnings, the principles have gone on to be used for biomedical ethics in general. They have been changed somewhat as the meaning of each is elaborated, as subdivision takes place, and as more principles are added (depending on the particular author). For example, for Beauchamp and Childress, the principle of beneficence spawns the principle of nonmaleficence. But, in whatever form, these principles of bioethics have come to dominate the field of bioethics. And that is why we will spend some time investigating several of them in detail.

Our general orientation toward the principles is that they express something very important, something very basic to our moral intuitions. However, for reasons to be seen, they are inadequate and misleading when used to resolve moral problems. So our plan is to show how our more comprehensive systematic account of morality can encompass and preserve what is good about the principles, while eliminating their unfortunate features. We see them as historically providing a conceptual ladder that allowed the field to achieve certain insights and goals. But, having arrived, the ladder is best set aside because it has become cumbersome and possibly dangerous.

Critique of Principlism

Our General Approach

Although we have been referring to principlism as a theory, it is really not so much a theory itself as a collection of "principles" that together are popularly thought to function as a theory in guiding action. Principlism emphasizes certain principles that it considers the "action guides" most relevant for use in resolving issues of biomedical ethics. One can find an assortment of principles claimed by various authors to be "the principles of biomedical ethics," but the ones most frequently and popularly found are those labeled "the principle of autonomy," "the principle of nonmaleficence," "the principle of beneficence," and "the principle of justice." Because these are the ones that are found most frequently together (and thus come to be taken as constituting a theory of biomedical ethics) and because these are the ones espoused by our paradigm progenitors of principlism, they are the ones we will analyze in order to contrast and compare with our own theory.

In the course of this chapter, we will show that principlism is mistaken about the nature of morality and is misleading as to the foundations of ethics. We will argue that its "principles" are really misnomers because, when seen for what they are, they are not action guides at all. Traditionally, principles really were action guides in that they summarized a whole theory and thus, in a shorthand manner, assisted a moral agent in making a moral decision. So, what then do the principles of principlism do that leads to their popular employment? We will argue that they primarily function as checklists naming issues worth remembering when one is considering a biomedical moral issue. "Consider this, consider that, remember to look for..." is what they tell the agent, rather than providing a guide about what to do in a particular situation. They clearly do not provide an explicit, systematic, and unified account of moral reasoning.

These principles presumably follow from several selected moral theories, though that connection is neither focused on explicitly nor clearly stated by the proponents of principlism. This becomes a matter of significant concern, because there seems to be no underlying connection between the principles. They do not grow out of a common foundation and they have no stated systematic relation among themselves. Although each may be an expression of one or another important and traditional concern of morality, there is no priority ranking among them or even any specified procedure for resolving cases of conflict that might arise between the principles. This serves to perpetuate what we call the "anthology syndrome." That is, it is a kind of relativism (perhaps unwittingly) espoused in many books (usually anthologies) of bioethics. They parade before the reader a variety of "theories" of ethics (Kantianism, deontology, Utilitarianism, other forms of consequentialism, and so forth) and say, in effect, choose whichever of the competing theories, maxims, principles, or rules that suits you. Similarly, the principles of principlism are unconnected with each other and, though each embodies a key concern from one or another theory of morality, there is no account of how they should relate to each other. Thus our conclusion will be that principlism obscures and confuses moral reasoning by its failure to provide genuine action guides and by its eclectic and unsystematic use of moral theory.

Our plan is to discuss very briefly the principles of nonmaleficence and justice in order to set the context for our argument. Then we will discuss in some detail the principles of autonomy and beneficence in order to demonstrate the force of our arguments against principlism. These latter two were chosen not only because they are the ones most often employed in discussions of biomedical ethics, but also because they best illustrate the most problematic aspects of principlism. In particular, we will show that principlism embodies the inadequacies of most previous accounts of morality by failing to appreciate the significance of the distinction between moral rules and moral ideals, by misrepresenting the ordinary concept of duty, and by failing to appreciate the importance of morality being a public system.

The Principle of Nonmaleficence

The reader will easily understand why this is the one principle toward which we would feel strong affinity. Chapter 2 makes clear that the key insight expressed by the principle of nonmaleficence is the same as that expressed by our account of morality.

This principle is the only one of the four principles that does not blur the distinction between moral rules and moral ideals. Indeed, this principle is most reasonably taken to be merely a summary of the first five moral rules. The moral rules "Do not kill," "Do not cause pain," and "Do not disable" are clearly included in this principle, and probably the rule "Do not deprive of pleasure" is as well. The rule "Do not deprive of freedom" also could be included in the principle of nonmaleficence, although the proponents of principlism seem to prefer to have it included under the principle of autonomy. However, we see no reason for distinguishing the rule concerning the deprivation of freedom from the other four rules, because all five of these rules prohibit causing what are universally recognized as harms or evils—that is, death, pain, disability, loss of freedom, and loss of pleasure.[4]

The principle of nonmaleficence, "Do not cause harm," at its best does no more than is done by our first five moral rules. That general principle "Primum non nocere" has been traditionally regarded as the first principle of medicine. It is primarily a matter of purpose and style whether one prefers to list five distinct moral rules or to have one general principle that includes them all. We prefer the former because it highlights the fact that there are different kinds of harms or evils and that rational persons can and do rank them differently. Neglecting the fact that there are different, but equally rational, rankings of the harms or evils is one of the primary causes of unjustified paternalism. Some people prefer to suffer pain rather than to lose freedom; some rank intense and continuing pain as worse than death itself; some believe a life of severe handicaps is not as bad as loss of life. There is no unique rational ranking of evils, and thus equally rational persons can often disagree. Imposing one's own ranking of evils on another is at the heart of paternalism. Thus, distinguishing between the different harms that one should avoid causing must be explicitly and carefully done in any event. So the gain in simplicity of having just one general principle rather than several distinct rules masks the underlying complexities. Nonetheless, this principle, even as it stands, has no major problems. That fact is not surprising, because it is the only one of the principles that is not an invention of philosophers but is a long-standing principle of medicine. However, because it is not incorporated into any public system, it is not clear when or how it should be overruled.

The Principle of Justice

Our discussion of justice will be equally brief, but not for the same reasons. Far from finding this principle as summarizing several specific moral rules, we find that this principle does not even pretend to provide a guide to action. It is doubtful that even the proponents of principlism put much stock in it as an action guide. The "principle of justice" is a prime example of a principle functioning simply as a checklist of moral concerns. It amounts to no more than saying that one should be concerned with matters of distribution; it recommends just or fair distribution without endorsing any particular account of justice or fairness. Thus, in the texts of principlism, the principle of justice, in effect, is merely a chapter heading under which one might find sophisticated discussions of various theories of justice. After reading such a chapter, one might be better informed and sensitive to the differing theories of justice but, when dealing with an actual problem of distribution, one would be baffled by the injunction to "apply the principle of justice."[5]

The principle of justice has an additional problem that it shares with the two remaining principles—it blurs the distinction between what is morally required (obeying the moral rules) and what is morally encouraged (following the moral ideals). Because the principle of justice can hardly be taken seriously as an action guide, this blurring is not as obvious as in the two remaining principles. In this, as in other matters, principlism simply takes over errors of those theories that suggested their principles in the first place. For example, the most prominent contemporary discussion of justice is by John Rawls in *A Theory of Justice*.[6] Rawls describes what he calls the duty of justice as follows:

> This duty requires us to support and to comply with just institutions that exist and apply to us. It also constrains us to further just arrangements not yet established, at least when this can be done without too much cost to ourselves.
>
> *(p. 115, see also p. 334)*

Thus Rawls includes in what he regards a single duty (1) the moral rule requiring one to obey (just) laws and (2) the moral ideal encouraging one to help make just laws, without even realizing how different these two guides to action are. As we shall see, this failure to distinguish between what is morally required (obeying the moral rules) and what is morally encouraged (following the moral ideals) also creates significant confusion in both the principle of autonomy and the principle of beneficence.

The Principle of Autonomy

This principle seems to be the centerpiece of principlism. It is cited more frequently than the others and has really taken on a life of its own. The concept of autonomy has come to dominate discussions of medical ethics—to the point that there is a growing and focused opposition to its predominance. Attention is being drawn to concerns that outweigh autonomy; it's claim of trumping every conflict of interests is being questioned.[7] But these developments are only symptomatic of deeper theoretical problems with autonomy as a principle. As close as Beauchamp and Childress get to stating the principle of autonomy is this:

> This principle can be stated in its negative form as follows: *Autonomous actions are not to be subjected to controlling constraints by others.* This principle provides the justificatory basis for the right to make autonomous decisions. The principle should be treated as a broad, abstract principle independent of restrictive or exceptive clauses such as "We must respect individuals' views and rights *so long as* their thoughts and actions do not seriously harm other persons." Like all moral principles, this principle has only prima facie standing. It asserts a right of noninterference and correlatively an obligation not to constrain autonomous actions.[8]

As stated here, it is surprisingly akin to the principle of nonmaleficence, and as such we would of course have little disagreement with it. In fact, it seems to pick out just one evil, the loss of freedom, and gives it a principle all to itself. If it were interpreted simply as an alternative formulation of our moral rule "Do not deprive of freedom," we obviously would have no objection to this principle. And it would be a real action guide in that it would tell one what to refrain from doing. However, the principle does not say simply that one should not constrain another's actions and choices, but rather it says that we should not constrain another's autonomous actions and choices. The addition of "autonomous" is what causes most of the problems with the principle of autonomy. That is, the principle does not prohibit the constraining of nonautonomous choices and actions. Consequently, this puts a great burden on the distinction between autonomous and nonautonomous choices and actions. What counts as an autonomous choice or action becomes a matter of fundamental moral concern.

Autonomous Actions and Choices In practice, the basic difficulty with autonomy, dogging it throughout all its uses, is knowing whether or not the actions and choices that one is concerned with are autonomous. Which is the autonomous choice: the decision to give up drinking or the decision to continue drinking? Is the choice to withdraw from expensive life-prolonging treatment to save one's family money and anguish the autonomous choice or is the autonomous choice to go on living a while longer? Which choice is it that we are being admonished not to constrain? This ambiguity invites a conflict between people who differ on which choice of the patient is the autonomous one. One side may favor overruling a patient's refusal on the ground that the refusal is irrational, claiming that therefore the choice is not autonomous; whereas the other side may favor going along with the patient's explicitly stated refusal on the ground that, though the refusal is irrational, the patient is competent and therefore the refusal is an autonomous choice. Both sides can claim that they are respecting the autonomous choice and, hence, acting on the principle of autonomy. A principle that can be used to support two completely opposing ways of acting even when there is no disagreement on the observable facts of the case is obviously not a very useful guide to action.

There may seem to be times when it is appropriate to question whether what the patient chooses is an autonomous choice—for example, when he is delirious, or intoxicated, or under the influence of drugs and the views he expresses significantly differ from those he expresses when he is in a normal state. But, even in these kinds of cases, it seems preferable to use the more commonly used concept of incompetence with regard to the patient rather than to appeal to autonomy. The only clear example of when one can say that the choice is not an autonomous choice is when the person is delirious or intoxicated and that condition results in a sudden change of view. But, even in this kind of situation, it is not always clear that one should not respect the patient's choice. Simply because a patient is incompetent, it does not follow that his choice should be overruled.

Even more importantly, when the significant departure from previously expressed views is not temporary and not explained by medical reasons, it is especially misleading and unhelpful to focus on the question of whether the patient's choices are autonomous. To do this is to substitute a metaphysical question for a moral one. There are no clear empirical criteria for the label's application. Thus, following the principle of autonomy may encourage one to act with unjustified paternalism—that is, to overrule the patient's explicit refusal simply because one views that choice as not being autonomous. Thus the principle of autonomy may lead one to deprive a person of freedom without an adequate justification for doing so.

A much more adequate method for dealing with such problems is by using the concepts of "rational" and "irrational." Only if a person's decision is seriously irrational are we justified in overruling it. If a person's decision for his own health care is rational, we are not justified in overruling it. Suppose, for example, that a patient had thoughtfully and persistently throughout his life said that if he ever had terminal cancer he would want no treatment at all. But now that he has cancer (and thus is anxious, stressed, and taking drugs for pain, and so forth) he says he wants life-prolonging treatment. Health care professionals would be hard pressed to decide what to do on the basis of whether or not this was an autonomous decision (after all, it was a sudden change of mind, under the influence of drugs and stress, and so forth). However, the patient's current decision is clearly not an irrational decision, and hence is not to be overridden.[9]

Moral Rules and Moral Ideals: A Fundamental Distinction At the core of many problems with the principle of autonomy is its failure in practice (and the failure of principlism generally) to recognize the significance of the distinction between what is morally encouraged (following the moral ideals) and what is morally required (obeying the moral rules). This distinction, or rather one that seems closely related to it, has traditionally been made by distinguishing between "duties of perfect obligation" and "duties of imperfect obligation" ("perfect" and "imperfect" "duties"). However, this indiscriminate use of the term "duty" (a matter we will discuss later in connection with beneficence) has made it almost impossible to make this crucial distinction in the correct way. All of the moral rules discussed in Chapter 2 would be counted as "perfect duties" and, unless one has an adequate justification for violating any of these rules, one is required to obey them. "Perfect duties" are those which one is required to impartially obey all of the time. One is morally allowed to violate a "perfect duty" only when one has an adequate justification for the violation.

On the other hand, the moral ideals would be counted as "imperfect duties"—that is, those "duties" that are impossible to obey either impartially or all of the time. Working to relieve pain would be an example. One may pick and choose not only which of those people suffering from pain to help, but also when and where one will provide this help. Furthermore, one may even choose not to act on that "imperfect duty" at all, but rather to act on some other "imperfect duty" such as preventing the deprivation of freedom of someone, somewhere. It seems as if an "imperfect duty" is a duty that one is not required to act on at all; morality certainly does not require one to work either for Oxfam or for Amnesty International, let alone both. It is not morally required to contribute to or work for any charity, although morality certainly encourages such behavior. Doing so is following an "imperfect duty" (moral ideal), not a "perfect duty" (moral rule).

Because this traditional distinction between "perfect and imperfect duties" embodies a confusion about the notion of duty, we make the distinction in a different and less misleading fashion. Moral rules prohibit causing or increasing the risk of harm and that is what morality *requires*. Moral ideals, on the other hand, admonish us to prevent harm, but morality can only encourage—not require—following those ideals. That is because, unlike the moral rules, a moral ideal requires a positive action to prevent or reduce harm, and hence it would be impossible to follow the moral ideals all the time, toward everyone, equally—that is to say, impartially. Doing what morality *requires* (i.e., obeying the moral rules) is not usually praiseworthy; rather obeying the moral rules is normally expected, and failing to do so makes one liable to punishment. Doing what morality encourages (i.e., following the moral ideals) is usually praiseworthy, but failing to do so is not punishable. It therefore is not surprising that it is basic and crucial to appreciate the distinction between the moral rules and moral ideals within the moral system.

Eliminating the misleading use of the terms "perfect and imperfect duties" makes clear that the distinction between moral rules and moral ideals is quite significant. The ordinary use of "duty" suggests that punishment or reprimand is deserved when one fails to do one's duty, perfect or imperfect. We are morally required to obey the moral rules impartially all of the time. All instances of killing, deceiving, cheating, and so forth, are immoral unless one has an adequate justification. For example, whenever one deprives persons of freedom (principlism might call this violating their autonomy), one needs an adequate justification for doing so. But one does not need a justification for failing to help them to increase their freedom (principlism might call this promoting their autonomy), unless one has a specific duty to do so—for example, because of one's profession. In the absence of such a duty, it is following a moral ideal to help someone to increase her freedom—it is something that morality certainly encourages us to do but not something that it requires us to do.

Autonomy as Rule and Ideal The principle of autonomy requires respect for autonomy, but it fails to distinguish between "respecting (not violating) autonomy" and "promoting autonomy." Not distinguishing clearly between "respecting autonomy" and "promoting autonomy" inevitably leads to confusion. Compounded by the search for the "genuinely" autonomous actions and choices, the principle of autonomy invites a kind of activism wherein an agent promotes those choices and actions of another that the agent regards as the other's autonomous choices and actions, even though that involves depriving that person of freedom. For example: Suppose a woman is pregnant with a fetus that tests have shown to be severely defective. The woman, believing she would like to have an abortion, consults a counselor. The counselor, knowing that the woman has always been "a good Catholic," sees her own duty to be that of dissuading the woman from an abortion. The counselor's reason would be that the decision to have an abortion would not be an autonomous choice. Such manipulation conflicts with morality itself insofar as it leads one to deprive people of freedom simply in order to promote what one decides would (or should) be their autonomous choice. Thus, principlism's centerpiece "principle of autonomy" embodies a deep and dangerous level of confusion. That confusion is created by unclarity as to what counts as autonomous actions and choices and the additional blurring of a basic moral distinction between moral rules and moral ideals. This unnecessary introduction of the metaphysical concept of autonomy inevitably results in making it more difficult to think clearly about moral problems. The goal of moral philosophy should be to clarify our moral thinking, not to introduce new and unnecessary complications.

As an aside, it is worth observing that the principle of autonomy probably caught on so tenaciously in the past three decades for two reasons. One is that Kantian ethics was experiencing a renaissance and that Kant's notion of autonomy was central to his account of morality. The other is that the medical profession had become so markedly paternalistic that patient self-determination had all but vanished. So the emphasis on autonomy became the banner under which patients' rights groups rallied to regain lost territory. Allowing the patient to decide what treatment he would receive became the main issue, and thus momentum and conviction—rather than conceptual clarity or theoretical soundness—perpetuated the emphasis on autonomy. Even the fact that the principle of autonomy did not really embody Kant's notion of autonomy did not detract from the overwhelming political appeal of invoking the principle.

An example of how confused the general understanding of autonomy is can be seen by examining Kant's view of autonomy. On Kant's view, one is not acting autonomously if one kills oneself or allows oneself to die because of intractable pain. That would be allowing pleasure and pain—which according to Kant are not part of the rational self—to determine one's actions. Thus such suicide (active or passive) would not be an autonomous action of the rational self. To act autonomously, one must always act in accord with the Categorical Imperative. In *The Grounding of the Metaphysics of Morals,* Kant explicitly states that the Categorical Imperative requires one not to commit suicide because of pain. By way of contrast, note that one of the major arguments in favor of allowing people to die when they are suffering from intractable pain is the principle of autonomy. The seeds of confusion were present in the initial planting of the concept of autonomy. This explains, in part, why we prefer the simple rule "Do not deprive of freedom" to the fancy principle of autonomy as a straightforward method of protecting patient self-determination.

The Principle of Beneficence

As used by principlism, this principle suffers shortcomings similar to those of autonomy. As popularly used in the biomedical ethics literature, this principle is cited presumably to validate both preventing or relieving harm and doing good or conferring benefits. Beauchamp and Childress, though much more cautious in their discussion of the principle of beneficence than many, do not avoid the errors. For them the principle of beneficence "asserts the duty to help others further their important and legitimate interests," and, being a duty, it is morally required. In the biomedical context, the principle becomes the duty to confer benefits and actively to prevent and remove harms, in addition to balancing the possible goods against the possible harms of an action.[10] Even though Beauchamp and Childress are well aware that many philosophers treat beneficent acts as "morally ideal," they still regard beneficence as morally required. But how could benefiting others ever be morally required of everyone? After all, as we have seen, impartiality is an essential feature of general moral requirements. But the general requirement of beneficence could never be impartially followed—that is, equally, toward everyone, all the time.

Thus the principle of beneficence not only succumbs to the same criticisms that we earlier leveled at the principle of autonomy for ignoring the distinction between the moral ideals (encouraging the prevention of harms) and the moral rules (prohibiting the causing of harms), but it is also subject to a new one—namely, failing to distinguish between the preventing or relieving of harm and the conferring of benefits (promoting goods). This distinction is especially important for medicine, inasmuch as preventing or relieving harm often justifies violating moral rules, whereas conferring benefits (or promoting goods) almost never does. For example, we might be justified in causing the pain of injection without informed consent if our purpose were to prevent some serious disease. But we would not be justified in causing the same pain without consent if it were in order to promote some good such as bulkier muscles or taller stature.

Beneficence and the Concept of Duty But there is another major confusion perpetuated by the principle of beneficence. Although the confusion arises from the mistake of turning moral ideals into duties, the problem itself goes beyond the rules/ideals distinction. Rather it involves a confusion concerning the concept of duty. Principlism considers it a duty to follow the principles. This especially becomes clear in the frequent references to "the duty of beneficence." It is not merely because it is a moral ideal that it is misleading to regard beneficence as a duty. (It would also be misleading to regard "Do not kill" as creating a duty, even though it is a moral rule.) Rather it is misleading because such usage distorts and obscures the primary meaning of "duty," which basically refers to the specific duties that come with one's role, occupation, or profession. Although it is correct to say "We ought not to kill" or "We ought to help relieve pain," it creates significant confusion to regard actions that we ought to do or refrain from doing as duties. For some philosophers, "Do your duty" has come to mean no more than "Do what you morally ought to do." But using the term "duty" in this way makes it very difficult to talk about real duties, those that are associated with one's profession and whose content is not determined by philosophers but by the members of that profession and the society in which they live. For reasons of conceptual soundness and clarity, we use the term "duty" only in its ordinary sense—that is, for what is required by one's role in society, particularly by one's profession. Thus not only is it misleading to talk of the moral ideals as imperfect duties, it is also misleading to talk of the moral rules as perfect duties.

Morality does put a limit on what can count as a duty: criminals have no duty imposed by their roles in a bank robbery, because there can be no duty to violate unjustifiably any of the moral rules. Nonetheless, "Do your duty" is a distinct moral rule on the same level as the other moral rules; it is not a meta-rule telling one to obey the other rules. It is justified as such because of the harm that would be caused by one's failure to do that which others are justifiably counting on being done. We are morally required to do our duty, but it clearly generates confusion to say that we have a duty to do our duty.

In medicine it is especially misleading to use the principle of beneficence as if it created a general duty for all health care workers. This obscures the role of real duties—that is, the specific duties that come with one's role or profession. Beauchamp and Childress seem to recognize the significant difference between what they call the general duty of beneficence and the specific duties of beneficence. They state: "Even if the general obligation of beneficence derives largely from reciprocity, specific obligations of beneficence often derive from special moral relationships with persons, frequently through institutional roles and contractual arrangements...."[11] They are clear that doctors, nurses, and others in the health care field have specific duties to their patients that are determined by their profession and by the practices of their specific institution. But to lump these varied and detailed professional duties together with the misconceived "general duty of beneficence" and place them all under one principle of beneficence is to cover up crucial distinctions.

Beneficence and Morality as a Public System Principlism fails to appreciate that morality is a "public system"—that is, that it must be known and understood by all moral agents and that it cannot be irrational for them to follow it. This failure to recognize that all justified violations of a moral rule must be part of a public system that applies to everyone, (that is, that everyone know that this kind of violation is allowed) is a crucial flaw. It may be helpful here to employ the distinction that was made in Chapter 2, between a moral system and a moral theory. A moral system is used to determine the morally right way to act; a moral theory is an attempt to justify that moral system—that is, to show that all rational people would favor adopting it as a public system that applies to everyone. Utilitarianism is sometimes presented as a moral theory that justifies our common morality, but often it is presented as providing an alternative to our common morality. As an alternative code of conduct, utilitarianism claims that we always ought to act so as to produce the best overall consequences. It is then appropriately criticized because it does not recognize the public nature of morality. Utilitarianism as an alternative code of conduct provides morally acceptable answers only when a way of acting that has the best consequences happens to coincide with that way of acting that is allowed by the moral system—that is, a public system that applies to all moral agents. Not surprisingly, these ways of acting often coincide, especially when considering those overt actions of government that can be publicly defended. This frequent but not constant coinciding explains both the seductive power of utilitarianism and why it is so easy to discover devastating counterexamples to it.

Because utilitarianism does not appreciate that all morally acceptable ways of acting must be publicly allowed, no principle derived from utilitarianism can be relied on to produce valid moral conclusions. The principle of beneficence, which is derived from utilitarianism, suffers most from this failure. Principlism, which does not recognize that all moral principles, including the utilitarian principle of beneficence, must always be part of a public system, therefore, cannot be trusted to yield morally acceptable decisions or judgments. When the consequences of violating a rule (for example, cheating) are very likely to be good in the particular situation but the consequences of publicly allowing that kind of violation would almost certainly be bad, principlism has serious problems. Because the principle of beneficence considers only the consequences, direct and indirect, of a particular act, it often encourages violating a moral rule even when the consequences of publicly allowing that kind of violation would be bad. It also seems to encourage causing great harm to one person in order to provide a lesser benefit to a sufficiently large number of others. Thus it is not surprising that the principle of beneficence often conflicts with the principles of nonmaleficence, justice, and autonomy. It is partly because the principle of beneficence sometimes leads to what everyone knows to be morally unacceptable conclusions that the principle of autonomy has attained such prominence in principlism. Without a full realization of what is happening, autonomy is used to overrule beneficence in many cases in which no rational person could publicly allow the kind of behavior that the principle of beneficence seems to require.

Autonomy has taken on such a prominent role in principlism because of principlism's failure to recognize the public character of morality. Recognition that all moral decisions and judgments must make reference to a public system that could be put forward by an impartial rational person would allow autonomy or, more clearly, freedom to take its proper place as one of many goods, not necessarily the highest of all goods. Freedom (autonomy) can be abridged if the harm being prevented to many is sufficiently great—that is, great enough that an impartial rational person could publicly allow that kind of violation. That limiting freedom to prevent harm can sometimes be publicly allowed explains why beneficence may sometimes overrule autonomy. Morality does not allow acting in any way that an impartial rational person could not publicly allow. But the principlism account of morality does not even recognize that all violations must be publicly allowed, thus it has no way of determining when autonomy overrules beneficence or vice versa. Principlism may therefore produce moral conclusions different from that which morality requires. This is most likely to happen whenever the conduct involved is beneficial only if not publicly known but would have serious negative consequences if everyone knew that this kind of behavior is allowed—for example, some cases of deception. In such cases, whether or not one favors acting on the principle of beneficence unfortunately seems to depend on the likelihood of exposure. In the absence of a recognition that any morally acceptable action must be publicly allowed, someone following the principle of beneficence may be led to accept acting in ways that clearly could not be part of any public system that would be put forward by any rational person.

This theoretical point is illustrated with regard to the issue of confidentiality of medical information. A breach of confidentiality, no matter how beneficial it may be in a particular situation provided that it does not become publicly known, often raises serious questions when evaluated in terms of its acceptability as a publicly allowed form of conduct. The principle of beneficence mistakenly suggests that our moral evaluation of such a breech should be significantly affected by our beliefs about how likely it is that this breech will be discovered and made public. However, morality requires considering the consequences of this kind of breech becoming part of the public system, allowable to everyone in morally similar circumstances—that is, sharing the same morally relevant features, even if it is very unlikely that this particular breech will be discovered. Not recognizing the public nature of morality, principlism sometimes simply claims that autonomy trumps beneficence and thus sometimes concludes that confidentiality be protected even when impartial rational persons could advocate publicly allowing this kind of violation. In certain circumstances, impartial rational persons may reasonably determine, when calculating the benefits and harms that would result from public knowledge of not protecting confidentiality, that the benefits would outweigh the adverse consequences. We should then adopt public policies that allow for breeches of confidentiality when the harms to be prevented are sufficiently great. As therapies and attitudes evolve, our changing policies toward the confidentiality of AIDS patients' information illustrates this point. For an additional example, see the next two chapters, where the question of whether to perform a medically unnecessary amniocentesis in order to protect confidentiality is discussed from the points of view of principlism (Chapter 4) and of our account of morality (Chapter 5).

Summary and Conclusion

The traditional concept of an ethical principle was one that embodied the moral theory that spawned it. As shorthand for the theory, the principle is used by itself to enunciate a meaningful directive for action. But it had an established, unified theory standing behind it. "Do that act which creates the greatest good for the greatest number." "Maximize the greatest amount of liberty compatible with a like liberty for all." The thrust of the directive is clear; its goal and intent are unambiguous. Of course there are often ambiguities and differing interpretations with respect to how the principle applies to a particular situation. But the principle itself is never used with other principles that are in conflict with it. Furthermore, if a genuine theory has more than one general principle, the relation between them is clearly stated, as in the case of Rawls's two principles of justice. Unlike principlism, Rawls does not present a number of conflicting principles, telling us to pick whichever combination we like.

The principles of principlism are used quite differently. In general they seem to function more as reminders of topics or concerns that the moral decision maker should review prior to decision. Except for the principle of nonmaleficence, they are not true action guides. The principle of justice is the clearest example of that. The principle of nonmaleficence is acceptable because it tells us simply not to cause harm, and as such merely summarizes our first four or five moral rules. Because it does not specify the harms that we are prohibited from causing, it is less useful than it might be. Because it does not make clear that there are different harms that different people will rank differently, it is more misleading than it might be. Nonetheless, insofar as the principle of nonmaleficence is interpreted as "Do not cause harm," it at least meets the criteria of being morally required.

The principles of autonomy and of beneficence are more complicated. They actually sound like action guides; they seem to tell one how to act. But on closer inspection we saw that they generated confusion. "Autonomous actions should not be constrained by others." If this were an action guide, like nonmaleficence, telling one what not to do—that is, Do not deprive of freedom—then of course we would have no objection because it is now synonymous with that moral rule. But unhappily the principle of autonomy goes beyond that clear and defensible rule. It injects the confusion over the metaphysical notion of "autonomous action" and, as we saw earlier, the principle upon further explication ends up requiring that we promote another's autonomy. That move, which fails to recognize the crucial moral distinction between a moral rule, "avoid causing harm," and a moral ideal, "prevent harm," means that the principle cannot be taken seriously as a moral requirement. If the principle is interpreted loosely to mean simply "respect persons" (the original principle from which the principle of autonomy seems to have been derived), it is still not clear what that would entail. At best it might mean, "Morality forbids you from treating others simply as you please. Some ways are acceptable and some are not. Think about it." So then we are back to the "list of concerns" interpretation of the principles.

The principle of beneficence also has that action guide appearance. It seems to be saying that one has a duty to prevent harm as well as to help others "further their important and legitimate interests." Besides the conceptual confusion over the notion of duty, this principle seems to be concerned primarily with what we call utilitarian ideals. It is certainly not a moral rule, because a person cannot possibly follow this principle impartially, all the time. It is not even a clear moral ideal, because the charge to confer benefits (unlike the charge to prevent harms) usually cannot be used to justify the violation of a moral rule. As in utilitarianism, from which this principle is derived, following it might lead to the unjustified transgression of moral rules toward a few in order to confer a benefit on the many.

And, lastly, we observed the more general difficulties with principlism. We noted that, even if the individual principles did contain action guides, they often conflicted with each other. Because they are not part of a public system, there is no agreed upon method for resolving these conflicts. Because they do not share a common ground, there is no underlying theory to appeal to for help in resolving conflicts. Indeed each of the principles, in effect, seems to be a surrogate for the theory from which it is derived. People use the principles in an unwitting effort to allow themselves to use whatever ethical theory seems to them best suited to the particular problem they are considering. It is simply a sophisticated technique for dealing with problems ad hoc.

The appeal of principlism is that it makes use of those features of each ethical theory that seem to have the most support. Thus, in proposing the principle of beneficence, it acknowledges that we should be concerned with consequences. In proposing the principle of justice, it acknowledges that we should be concerned with the distribution of goods. In proposing the principle of autonomy, it acknowledges that we should emphasize the importance of the individual person. In proposing the principle of nonmaleficence, it acknowledges that we should emphasize the importance of avoiding harming others. There is no attempt, however, to see how these different concerns can be blended together as integrated parts of a single adequate theory, rather than being disparate concerns derived from several competing theories. So in effect principlism tells agents to pick and choose as they see fit, as if one could sometimes be a Kantian and sometimes a Utilitarian and sometimes something else, without worrying whether the theory that one is using is adequate or not. Principlism does not recognize that any moral decision that one makes must be a moral decision that one could publicly allow. Principlism not only does not recognize the unified and systematic nature of morality, it does not recognize that morality (that is, the moral system) must be public.

The upshot of having principles with an unclear content that are not part of any unified public system is that an agent will not be aware of the real grounds for his moral decision. Because the principle is not a clear, direct imperative at all, but simply a collection of suggestions and observations, occasionally conflicting, the agent will not know what is really guiding his action. Without an account of the morally relevant features, he will not know what facts are morally relevant, such that a change in them may change what he should do. Although the language of principlism suggests that he has applied a principle that is morally well established, a closer look shows that in fact he has looked at and weighed many diverse moral considerations, which are only superficially interrelated, having no unified, systematic underlying foundation. Principles seem to be involved in complex decisions only in a purely verbal way; the real guiding influences on the moral decision are not the ones the agent believes them to be. Rather, the agent is, in fact, guided by his basic understanding of common morality and simply uses principles when stating his conclusions.

Endnotes

1. Beauchamp, T.L. and Childress, J. *Principles of Biomedical Ethics*, 3d ed., Oxford University Press, New York, 1989.
2. *Federal Register*, vol. 44, no. 6, April 18, 1979, p. 23192.

3. Op. cit., p. 23193.

4. Beauchamp and Childress, pp. 122–125.

5. *See* Clouser, K.D., and Gert, B., A critique of principlism, *The Journal of Medicine and Philosophy* 15: 225–227 (April 1990); and Morality vs principlism in Raaman Gillion (ed.), *Principles of Health Care Ethics,* John Wiley, 1993.

6. Rawls, J. *A Theory of Justice.* Harvard University Press, 1971. For further discussion of Rawls on this point, *see* Bernard Gert, *Morality,* Oxford University Press, New York, 1988, chapter 13.

7. It is to the credit of Beauchamp and Childress that they make it clear that other considerations can outweigh autonomy, pp. 112–113.

8. P. 72, their emphasis.

9. *See* section "Rationality as Avoiding Harms" in Chapter 2 of this book. *Also see* Charles M. Culver and Bernard Gert, The inadequacy of incompetence, *The Milbank Quarterly*, vol. 68, no. 4, 1990, pp. 619–643.

10. Beauchamp and Childress, pp. 194–195.

11. Beauchamp and Childress, p. 204.

4

Ethical and Legal Dilemmas Arising during Predictive Testing for Adult-Onset Disease: The Experience of Huntington Disease

The goal of predictive testing is to modify the risk for currently healthy individuals to develop a genetic disease in the future. Such testing using polymorphic DNA markers has had major application in Huntington disease. The Canadian Collaborative Study of Predictive Testing for Huntington Disease has been guided by major principles of medical ethics, including autonomy, beneficence, confidentiality, and justice. Numerous ethical and legal dilemmas have arisen in this program, challenging these principles and occasionally casting them into conflict. The present report describes these dilemmas and offers our approach to resolving them. These issues will have relevance to predictive-testing programs for other adult-onset disorders.

Introduction

DNA testing for diseases with onset in childhood is in many instances synonymous with diagnosis of disease. The new dimension to DNA testing for adult-onset disorders is that such testing may predate onset of symptoms by many years. Healthy individuals at high or low risk of developing a disease at some time in the future may be identified. The use

Note: This chapter was written by Marlene Huggins, Maurice Bloch, Shelin Kanani, Oliver W. J. Quarrell, Jane Theilman, Amy Hedrick, Bernard Dickens, Abbyann Lynch, and Michael Hayden. Reprinted with permission from The University of Chicago Press from Huggins, M. et al. *Am. J. Hum. Genet.* 47:4–12, 1990.

of DNA markers for prediction of future risk has had major application for Huntington disease (HD).

HD is an inherited autosomal dominant neuropsychiatric condition which usually manifests in midlife with chorea, cognitive impairment, and affective disturbance (Hayden 1981). The discovery of linked polymorphic DNA markers for HD has led to the development of predictive-testing programs which modify the likelihood that persons at risk for HD have or have not inherited the HD gene (Gusella et al. 1983; Hayden et al. 1988; Meissen et al. 1988; Wasmuth et al. 1988; Whaley et al. 1988; Brandt et al. 1989). Treatment to alter the natural history of HD is unavailable. In contrast, presymptomatic diagnosis of other late-onset disorders, such as familial hypercholesterolemia or polycystic kidney disease, may allow the adoption of preventive therapy which may alter the severity of the disease.

Potential ethical issues associated with predictive testing for HD were raised prior to the implementation of testing programs (Craufurd and Harris 1986; Lamport 1987; Shaw 1987). The Canadian Collaborative Study of Predictive Testing for Huntington Disease has been guided by four ethical principles which are essential to all clinical practice—namely, autonomy, beneficence, confidentiality, and justice.

The principle of autonomy requires respect for the individual's right to make an informed decision about an action which may have a profound effect on his or her life. Autonomy has been proposed as the primary moral principle which takes precedence over all other considerations (Beauchamp 1982). Two conditions of this principle are freedom from coercion and full understanding of the implications of that action. The physicians and counselors have an obligation to provide up-to-date information to the participants such that they are fully informed regarding all aspects of testing. At the same time, the counselor has a responsibility to assess whether a decision is being unduly influenced by other persons or institutions. The principles of confidentiality and privacy are derivatives of autonomy.

Beneficence is a fundamental principle of medicine, one which is summarized by the phrase "primum non nocere," or "first do no harm." This includes four elements: not inflicting harm, preventing harm, removing harm, and doing good. This implies not only avoiding harm to the patient or client but also preventing harm to other identifiable individuals. A broad analysis of beneficence is particularly important for predictive testing for HD which involves many family members. The principle of beneficence especially underlies the responsibility of genetic counselors to avoid harming persons who are not psychologically or emotionally equipped to safely deal with predictive-testing results. One of the fundamental goals of predictive-testing programs for HD is to document the long-term outcome for those persons receiving a modified risk, since beneficence has not yet been fully assessed.

Confidentiality requires particularly careful attention when one is dealing with predictive testing on related individuals, if inadvertent disclosure of information to third parties, including family members, is to be avoided.

The principle of justice implies fairness, which requires equal access to health services. The introduction of the Canadian Collaborative Study in 1988 has ensured that predictive testing is available throughout the country. A major priority of this program is to obtain long-term support and to develop guidelines for the continuation of the testing as a medical service so that this information, if desired, is not unreasonably difficult to obtain. In addition, the principle of justice underlies the right of the medical profession to decline to perform unnecessary tests or procedures and to avoid wasting resources.

Numerous ethical and legal dilemmas have arisen in our predictive-testing program, challenging the principles of autonomy, beneficence, confidentiality, and justice. The present report describes some of these dilemmas and presents our approach to resolving these problems. Many of these issues are not unique to HD and will have relevance to predictive-testing programs for other adult-onset disorders.

Case Studies

Case I: Issues of Confidentiality (Figure 4.1)

Two siblings requested predictive testing. The probands' mother, her maternal cousin, and the maternal grandmother were affected with HD. DNA analysis revealed that proband III-2 is at high risk (97%) of having inherited the gene for HD, while proband III-1 has a low risk (approximately 3%).

In this family the brother and sister were seen together in the early stages of the program, so each clearly was aware that the other was having the test. Person III-2 later elected not to inform his sister of his own increased risk result and instead told her there had been some delays in the laboratory and that his results were not yet available. The sister, person III-1, at the time she received her own results, expressed regret that her brother had not yet been able to have any news from the test and enquired when he might receive his results. The counseling team was

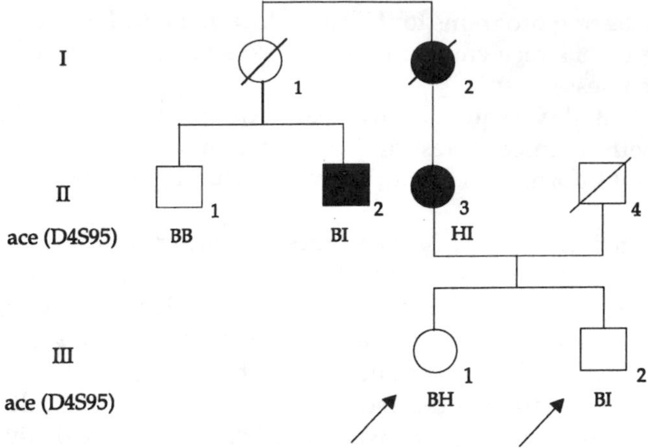

Figure 4.1 Candidates III-1 and III-2, who have inherited different haplotypes from their affected mother (II-3). For example, candidate III-2 inherited the I allele and has a high risk for developing HD. Candidate III-1 inherited the H allele from her mother and has a very low risk of developing HD.

aware that the brother had been given results and that he obviously had not informed his sister of his results.

Ethical Perspective The HD status of III-1 is not affected by knowledge of the HD status of III-2. Person III-1 should be reminded of this and of the program's emphasis on confidentiality. The natural affectionate and helpful concern of one sibling for another can be supported, but there is no ethical requirement to disclose either the sibling's result or the fact that he had been given results. On the contrary, there is an ethical obligation to respect the confidentiality of both probands and, therefore, to withhold from proband III-1 any information which belongs to proband III-2. Confidentiality could theoretically be overridden, if withholding of the information would result in significant and serious harm (United States President's Commission for the Study of Ethical Problems in Medicine and Biomedical and Behavioral Research 1983). There are no compelling reasons (i.e., prevention of harm) to override confidentiality under the circumstances in this situation.

Legal Perspective The patient's legal entitlement to confidentiality cannot be violated. The sister can only be informed that, because of her

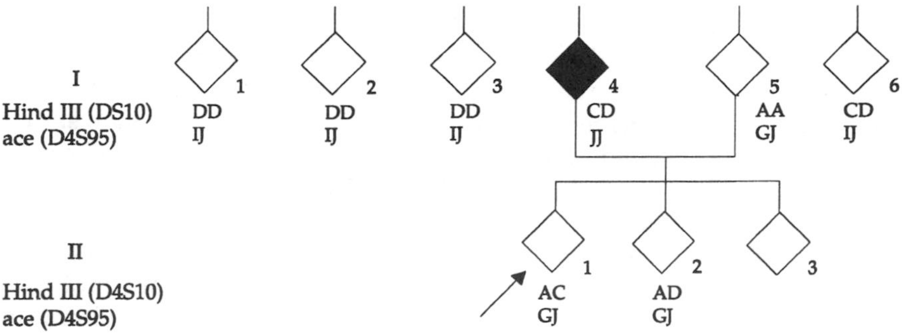

Figure 4.2 Proband II-1, who inherited haplotype CJ from the affected parent and was given an increased risk for HD. Sibling II-2 inherited a different haplotype (DJ) from the affected parent.

brother's control of any information he may receive about himself, he will be her sole source of that information.

Resolution Care was taken not to inadvertently disclose that the brother had been given his results. His decision to withhold the information was respected. It was important to allow this test candidate whatever time he required to come to terms with his change in status with regard to HD. In due course he did inform his sister of his test results, and both of them were able to deal effectively with their altered status. This illustrates both the need for care when one has more than one participant in a family and the importance of respecting the rights of all of the individuals concerned.

Case 2: Unconfirmed Diagnosis in a Family Member (Figure 4.2)

The proband (II-1) presented for predictive testing and was given a risk of 80% of having inherited the gene. This was a partially informative result because of the availability of DNA from only a few family members, including one affected person. Thereafter we learned from relatives that the sibling of the proband was probably showing early signs of HD. This was communicated to us independently on a few occasions by family members. However, this person had not consulted a physician for many years and had never been formally diagnosed as having HD. This individual had previously given blood and consent for the DNA analysis which was to be used to reconstruct haplotypes of the affected parent. Even though he knew he could participate in the program, he did not want any modification of his own risk relative to HD but gave blood, which might increase the informativeness for other relatives. The DNA analysis revealed that the proband and his sibling had inherited different chromosomes 4 from their affected parent (Figure 4.2).

If the sibling of the proband indeed has HD, then the proband would have a very low risk of developing HD, contary to what he had been told. In order to maximize the predictive-testing information for our proband, suggestions were made by family members to encourage the sibling to have medical intervention.

Ethical Perspective In view of the fact that privacy and confidentiality are primarily ethical considerations, the sibling cannot be asked to undergo clinical assessment. The psychological assessment of the proband did not suggest any compelling need for overriding the sibling's rights of confidentiality and privacy in order to potentially reduce the risk of harm to the proband. Pressure on the sibling to undergo clinical examination could foreseeably have increased significant psychological stress on him, thereby violating the principle of doing no harm to the sibling. Thus the proband will have less than optimal information regarding his own risk. In order not to interfere with the sibling's autonomy, the proband could only be reminded that the result given to him incorporates only the information about relatives whose current status with respect to HD is definitively known. This approach gives primacy to the sibling's rights of autonomy and confidentiality and is in accord with the duty to do no harm to the sibling, but it acknowledges, to a lesser extent, beneficence as it applies to the proband.

Legal Perspective The legal right to control information includes the right to select not to receive and to not actively seek to acquire it. Informed choice also includes informed refusal (*Truman v. Thomas* [1980]) to participate in testing.

Resolution The status of the sibling with respect to HD is regarded as uncertain. We offered our services for assessment and counseling should this person decide to seek medical attention at some time in the future. The proband was informed that, if any other family member develops HD, this may significantly alter the estimate of his or her risk.

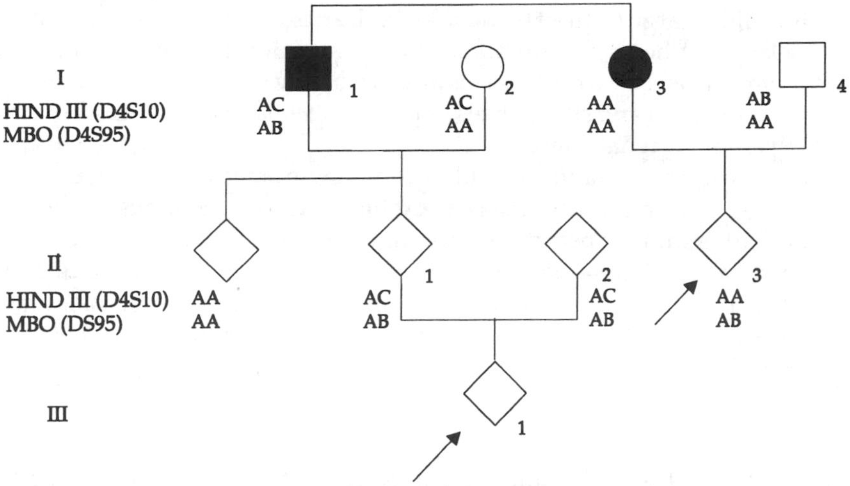

Figure 4.3 Proband II-1, who has requested prenatal exclusion testing. First-cousin II-3 has concurrently requested predictive testing. Candidate II-1 has serendipitously been shown to have a low risk of developing HD.

Case 3: The Problem of Too Much Information (Figure 4.3)

Person II-1 has requested prenatal exclusion testing and would prefer not to have any modification of his or her own risk for HD. This type of testing has been described elsewhere (Quarrell et al. 1987; Fahy et al. 1989) and could certainly be done in this family. DNA analysis would include the parents (II-1 and II-2) of the fetus and the affected grandparent (I-1).

Concurrently and without the knowledge of person II-1, a relative (II-3) requested predictive testing. In order to undertake the analysis, we obtained blood from additional crucial relatives in this family. This resulted in information that person II-1 is at low risk of developing HD, and prenatal testing could be considered unwarranted. However, proband II-1 has requested prenatal exclusion testing; the express wish was to receive no alteration in his or her own risk for developing HD. This would require that we ignore the knowledge gained serendipitously and that we provide only the test which the candidate requested, resulting in prenatal testing on a fetus which is known to be at low risk.

It could be argued that it would be in the best interest of the candidate to disclose that he or she has a very low risk of developing HD and that further testing is unwarranted. This approach may significantly reduce the proband's anxiety and stress levels, and it also avoids the unnecessary risk and expense of prenatal testing.

However, the situation would become increasingly complex if other family members requested prenatal exclusion testing. Siblings of person II-1 and others might become aware that we have a policy of disclosing serendipitously discovered "low risk" results to prevent an unnecessary prenatal procedure. Any siblings who are not told that the prenatal test is unwarranted might infer that their own risk of having inherited the gene for HD is high.

Ethical Perspective The principle of beneficence supports disclosing to person II-1 his low risk status, since this approach might significantly reduce his anxiety level. However, beneficence also requires that we avoid harm to other at-risk HD family members who may foreseeably be adversely affected by making unwarranted inferences as a result of a policy of disclosing serendipitously discovered low risk results to avoid unnecessary prenatal procedures. Any participants who are not told that the prenatal test is unwarranted might infer, with potential adverse consequences, that their own risk for HD is high. In addition, following disclosure of the low risk results, person II-1 might realize that a family member had had predictive testing. Not inflicting harm must take priority over doing or promoting good (Beauchamp 1982).

Concerns for both noninjury/beneficence for the fetus and justice (nonwaste) for society can be considered secondary to protection of confidentiality. Discussion of the risks to the fetus of any prenatal test is essential, but the negative consequences of revealing to person II-1 any information personal to other family members strengthens the argument for nondisclosure. In addition, as long as person II-I is aware of other options for predictive testing and has made an informed decision not to proceed with definitive testing for him- or herself, there is an ethical obligation to respect that autonomous decision.

Legal Perspective It is legitimate to perform the requested test in these circumstances, unless the test itself presents undue risk to either the parent or the fetus. The counselors are bound by the legal constraints of privacy and confidentiality, since information personal to other family members

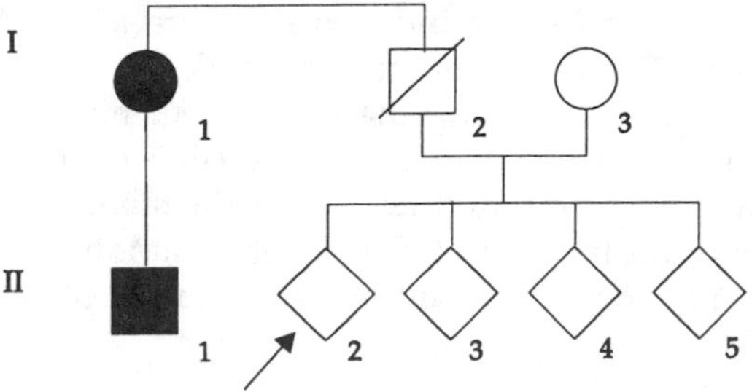

Figure 4.4 Proband II-2, at 25% risk for HD, who requested predictive testing. The father (I-2), died at age 36 years, with no signs of HD. Definitive testing would be possible if DNA analysis was done on the proband's siblings in an effort to reconstruct the haplotypes of the father.

cannot be disclosed without compelling reasons (such as avoiding undue risks or harm) to override confidentiality.

Resolution Our approach in this situation is to do the prenatal testing which the candidate has requested but to give a low risk result whatever DNA marker is inherited by the fetus. In pretest counseling we would discuss with person II-1 the possible outcomes of the prenatal exclusion test. Candidates undergoing this form of testing are told either that the fetus has a very low risk for HD (i.e., that we have been able to "exclude" HD) or that the risk is increased and similar to that for the affected parent. We would give no information from which one might draw conclusions about the status of the at-risk parent who does not want definitive testing.

We have respected the confidentiality of both candidates and have been careful not to reveal that these two related individuals are having testing concurrently. We have upheld the principles of autonomy and privacy and provided only the information which has been requested. In so doing, however, we have performed a prenatal diagnostic procedure on a fetus serendipitously known to be at low risk.

Case 4: Reconstructing Haplotypes (Figure 4.4)

In this family the proband's father (I-2) died at age 36 years, with no clinical signs of HD, and it is not known whether he had inherited the HD gene. There is a documented family history of HD in that the paternal grandmother and several other individuals were affected. In order to provide definitive information for the proband, the father's haplotypes could only be reconstructed using the DNA from his children (persons II-2, II-3, II-4, and II-5). However, the proband (II-2) is the only individual in this sibship who has requested testing. If the siblings' DNA samples are analyzed for reconstruction of the father's haplotype, then information could be learned regarding their individual risks of having inherited the HD gene.

Ethical Perspective Ethically, no principle is violated by having the information about the siblings' risk known to the laboratory, as long as aspects of confidentiality are absolutely adhered to. This testing should only take place if those people donating blood are aware that their samples may be analyzed to aid in predictive testing for relatives. Individual rights to privacy and rights not to know can be protected.

Legal Perspective Living individuals have a legal right not to have uninvited persons acquiring medical/genetic information about them. These rights may be protected by disclosing the outcome for the proband in a way which precludes anyone learning information about other individuals at risk. In addition, the right to privacy may be protected by ensuring that those performing the tests are unaware of the identities of all subjects except those who are seeking information.

Resolution In this particular family we began by using an exclusion testing approach, whereby it was not necessary to analyze the siblings' DNA. We demonstrated that the proband shared no haplotype in common with the affected aunt and cousin. Thus, no further testing was required, and the proband was given a risk of 3% of having inherited the HD gene. This case emphasizes both the importance of informed consent and the need to limit DNA analysis to those persons whose DNA must be included if an informative result is to be reached.

Our approach in this program has been to utilize DNA samples that are available, in order to provide an informative result for the proband, but, whenever possible, to take precautions to learn nothing about the status of other at-risk individuals. When siblings are needed for haplotype analysis, we take DNA from the siblings (if more than one) but place it in tubes marked (e.g., "A," "B," or "C"). No record is kept of which sibling's DNA is in tube A or B or C. At no time does the person doing the DNA analysis know which DNA sample corresponds to which sibling. The results are used to reconstruct the father's DNA haplotypes without learning the individual risks of the siblings.

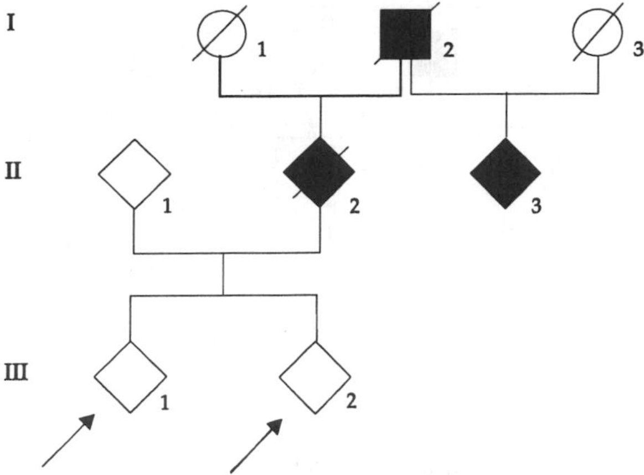

Figure 4.5 Two siblings who have requested predictive testing. Their affected parent is deceased, as are the grandparents. No blood samples could be obtained from the only living relative with HD (II-3).

Case 5: Inability to Obtain Blood from a Crucial Relative (Figure 4.5)

In this family the affected mother (II-2) is deceased, and the only living relative with HD is her half-sibling (II-3), who is elderly and severely incapacitated with HD and who has been deemed an incompetent person. Although the predictive test is unlikely to be highly informative in this family, it may be possible to provide at least some alteration in risk for those individuals requesting the predictive test. However, all efforts to obtain consent for blood withdrawal from the affected individual have failed, because of longstanding conflicts within the family and because of failure to get permission from the appointed guardian. Without a sample of DNA from this crucial relative, it is impossible to offer predictive testing for probands III-1 and III-2. The at-risk individuals have been investigating options open to them with regard to petitioning the courts to request the cooperation of the affected relative.

Ethical Perspective The priority of individual autonomy over non-injury/beneficence to others provides the ethical basis for noncoercion of person II-3 in the matter of consent for blood withdrawal.

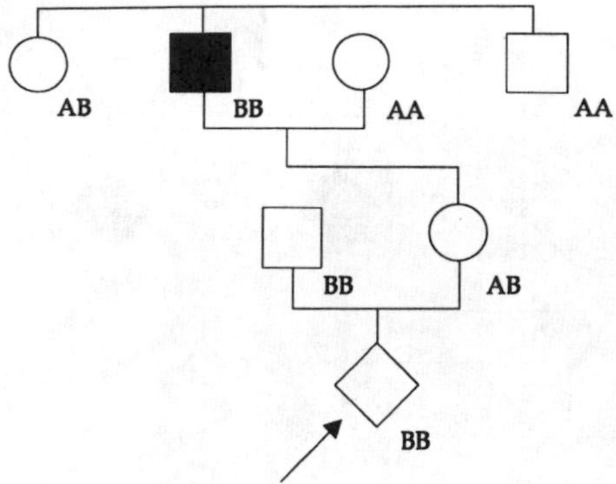

Figure 4.6 An at-risk pregnant mother who inherited marker A from her mother and marker B from her affected father. The fetus was homozygous BB, indicating that it inherited the B allele from its maternal grandfather. Prenatal exclusion testing increased, from 25% to 48%, the likelihood that the fetus inherited the gene for HD (Hayden et al. 1987).

Legal Perspective It is reasonably clear that Canadian and American courts reluctant to approve blood testing of unwilling competent persons for their own benefit will not generally authorize this invasive procedure on an incompetent person for the benefit of another (*McFall v. Shizp* [1979]). Venipuncture—the breaking of the outer surface of the skin—is in law more than an assault, which is an unauthorized touching. It is a wound, and courts will not compel one person to be wounded for the possible advantage of another.

Resolution Predictive testing became possible in this family when another relative was diagnosed with HD. It was no longer necessary to obtain blood from the aunt. Legal issues with respect to predictive testing remain untested in Canadian courts.

Case 6: Problems of Terminating a Fetus at 50 Percent Risk of Developing a Late-Onset Illness (Figure 4.6)

A woman in her early 20s requested prenatal diagnosis but did not wish to have any alteration in her knowledge of her own risk for HD. In this family it was possible to clearly distinguish the maternal grandmother's marker from that of the affected maternal grandfather. The mother inherited the marker A from her mother and inherited marker B from her affected father. The fetus proved to have inherited marker B, implying that the fetus inherited a chromosome 4 from its affected grandfather. The likelihood that the fetus has inherited the gene for HD rose from 25% to close to 50%. The pregnancy was terminated at 13 week gestation (Hayden et al. 1987).

Subsequently, there has been considerable ongoing contact with other family members, with much discussion centering around whether prenatal exclusion testing should have been offered. The unaffected parent of our proband has expressed concern for the future implications of having terminated a pregnancy at 50% risk for a late-onset illness. She has requested that such testing be avoided.

Ethical Perspective There was an ethical obligation to ensure that the proband understood that she could have had better knowledge on which to base her decision about termination of the pregnancy. If she then elected not to acquire such knowledge, her decision to have no personal predictive testing remains her own, as does the responsibility for the consequences of that decision. The respect for autonomy demands that individuals at risk for HD be free to make their own informed decisions regarding testing, irrespective of the objections of others.

Legal Perspective In the absence of criminal law restricting abortion, the legal question of whether abortion constitutes medical professional misconduct relates to whether it is ethically discussed and undertaken. Under the doctrine of informed consent, it is essential that the proband be thoroughly advised about the limits of prenatal testing, as well as about all alternatives in predictive testing. The rights of the pregnant woman take precedence as long as the fetus is not viable. Individuals are free to take action which they feel is appropriate to their circumstances. There is no legal impediment to exclusion testing.

Resolution Prenatal exclusion testing is desired by some at-risk individuals who prefer not to have any modification of their own status with regard to HD. At the present time, this test is one of several options open to candidates in our study. These options are fully discussed, and each candidate makes the decision which is appropriate for him or her at the time he or she enters the study, according to the principle of autonomy.

Case 7: Rights of Individual versus Rights of Society

We have had a request from a major air transportation company to do the predictive test on an employee without informing the latter specifically, by obtaining a blood sample under false pretenses. This at-risk employee is a pilot, and the company wished to learn whether he is likely to have HD in the future, because it could influence his continued employment as a pilot. This pilot, at the time of the company's request, did not wish to have predictive testing because he felt that the information could be detrimental to his future career plans.

Ethical Perspective To act in a fashion which is deceptive would be ethically wrong; to break confidentiality would be contrary to the ethical priorities of the predictive-testing program. However, beneficence requires that harm be prevented to members of society who are affected by the air company's transportation service. This principle can be honored by regular medical examination. Public safety can be protected without predictive testing being done.

Legal Perspective The employer's proposal is unlawful. The Supreme Court of Canada held in December 1988 that physical sampling cannot be undertaken for purposes of obtaining identified information about a person for which his or her prior consent has not been given. While in the United States there is no Supreme Court decision, there is a small body of case law to support this principle (*Hague v. Williams* [1962]).

Resolution This request was denied. It was suggested that the employer discuss this matter further with the at-risk individual. At present, predictive testing cannot be done without blood samples from numerous family members. However, when the gene is cloned, it may be possible to test a single sample. Nevertheless, the employee would be tested only with his or her express voluntary consent, according to the principle of autonomy.

Case 8: Predictive Testing for MZ Twins

One twin of a presumed MZ twin pair at risk for HD requested predictive testing. Any results given to one twin would also apply to the co-twin, who did not desire to have any alteration in his or her knowledge of their risk to develop HD. To accede to the wishes of one twin is to deny the rights of the other twin.

Ethical Perspective In this instance there are two persons being tested—one who is receiving counseling and support and one who has not even entered the program. The proband has a right to know his or her own status, according to the principle of autonomy. At the same time, the co-twin has both a right to privacy and a right to decline to have information. In addition, the issue of beneficence cannot be ignored; testing of the co-twin without obtaining informed consent and without having appropriate support in place may pose significant risk to this person.

Legal Perspective The legal requirement of obtaining informed consent to testing is reinforced by the requirement that any consent be freely given. There is no legal obstacle to testing the one twin. However, there is also no impediment to our program adopting a protocol that stipulates that testing one twin is dependent on the requirement that the other twin give informed consent.

Resolution Although there is no legal objection, in our program any testing of one MZ twin without obtaining informed consent of the co-twin would be denied because of clinical and ethical concerns. In this instance, beneficence must take priority over autonomy. Joint consultation with both twins may be helpful in resolving this conflict. With regard to this particular family, we learned that the twins were not MZ, and thus testing for the proband proceeded in the usual fashion.

Case 9: Knowledge of Unsolicited Information

A man at risk for HD and his spouse have requested prenatal exclusion testing. He declined to have definitive testing for himself. DNA typing done prior to the prenatal test to identify which markers are informative revealed that the affected parent was not the biological father of the proband. That is, a person who requested prenatal testing because he was at risk for HD was discovered not to be at risk for HD. Should prenatal testing for HD be done in this family when we know that the at-risk parent is in fact not at risk for HD?

Ethical Perspective There is an ethical responsibility to fully inform candidates of possible outcomes, including possible nonpaternity in cases of putative patrilineal HD. This responsibility extends to persons donating, for DNA banking, blood which may be used for predictive testing in the future. The possibility of discerning nonpaternity through DNA testing should be explicitly covered in the informed consent for both DNA banking and predictive testing. Respect for autonomy demands that individuals be allowed to say whether they would want to be told of nonpaternity. In the prenatal context, one could argue in favor of disclosure so that an unwarranted prenatal procedure could be avoided. This approach, however, infringes on the privacy of the proband's parents and allows the principle of justice for society to take precedence over the principle of autonomy. However, it has already been stated that concerns for both noninjury to the fetus and justice for society can be considered secondary to autonomy in this predictive-testing program. Therefore nonpaternity should be disclosed only if the proband expresses his wish to be fully informed, according to the principle of autonomy.

Legal Perspective The physician's duty to give material information to a patient prior to undertaking a procedure requires disclosure of whether a test will yield information not already available. The depth of explanation regarding possible unsolicited information depends on the circumstances, including the patient's response to questions regarding how much explanation is desired.

Resolution This situation is not unique to HD. Our approach is to disclose the information regarding nonpaternity if the proband expresses a wish that this be done. No inference can be made about any other at-risk person, and an unnecessary prenatal procedure will be avoided. If the proband declines to be fully informed, we will do the testing which has been requested but will give very low risk results (as in case 3). The primary principle to be followed here is to provide accurate information of the genetic risk to the proband. In this particular instance, the pregnancy ended in spontaneous abortion at 10 wk gestation, prior to any intervention. The proband has not returned to the program.

Discussion

Predictive testing for HD will eventually serve as a model for the application of predictive DNA tests for other late-onset genetic diseases. Many of the ethical dilemmas presented here are not unique to HD, but they have demonstrated some of the difficulties which must be addressed before this type of testing becomes widely available.

Despite years of careful planning, predictive testing is far more complex than had been anticipated. Flexibility in the guidelines for participation has been necessary, and a team approach to resolve the difficulties has been helpful. In difficult situations, we have presented the dilemmas to an advisory board including a geneticist, an ethicist, and a lawyer, as well as representatives from the local HD society who serve to provide support and guidance in resolving these problems.

We have learned the importance of both listening carefully to the precise requests of participants and exploring with them their reasons for having testing. Certain candidates want some pieces of information and not others. For example, some people participating in predictive testing do not want to know the result of their physical examination. In other words, they are prepared to learn about future risk but not to learn about possible onset of disease. For this reason, we always ask whether participants wish to learn the results of the clinical examination.

One of the major ethical principles which has guided the development of our program has been the principle of autonomy; each at-risk person has the right to decide whether to participate in predictive testing. We aim to provide to participants the most accurate information, including a discussion of different options for predictive and prenatal testing. A detailed informed consent dealing with possible outcomes is crucial prior to a widespread implementation of such programs.

As a general rule, autonomy has taken precedence over other issues, and we have considered that our primary allegiance has been toward the candidates presenting for information important to their health. However, in certain situations the right of the individual to have this genetic information has been in direct conflict with the rights of family members not to know. Thus, in the case of MZ twins, autonomy is superseded by significant concerns of possible harm to the co-twin, and testing would be denied in the absence of mutual consent.

In some situations, in the interests of beneficence a more directive approach to counseling is appropriate. For example, there have been a few instances in our program when detailed pretest assessment and counseling have indicated areas of significant concern, such as a past history of suicide attempts, current depression, or lack of social supports. In these situations immediate continuation of testing has been judged to pose undue risk to the participants; thus, testing has been postponed in favor of more detailed psychiatric assessment and counseling.

This and similar programs have led to the designation of a new category of at-risk person in families with HD—namely, the presymptomatic person (Wexler 1989). Careful long-term assessment and documentation of the impact of such testing on all participants and their families is needed so that guidelines can be developed to ensure appropriate use of predictive tests in both HD and other late-onset genetic diseases.

Acknowledgments

We thank our colleagues in the Canadian Collaborative Study of Predictive Testing for Huntington Disease. The comments of the reviewers were particularly helpful. This work is supported by the MRC (Canada), the National Health Research Development Program, and a western Canadian private foundation. Michael Hayden is an investigator of the British Columbia Children's Hospital.

References

Beauchamp, T.L. (1982). Ethical theory and bioethics. In: Beauchamp, T.L., Walters, L. (eds). *Contemporary Issues in Bioethics*, 2d ed. Wadsworth, Belmont, California, pp. 1–43

Brandt, J., Quaid, K.A., Folstein, S.E., Garber, P., Maestri, N.E., Abbott, M.H., Slavney, P.R., et al. (1989). Presymptomatic diagnosis of delayed-onset disease with linked DNA markers: the experience in Huntington's disease. *JAMA* 261: 3108-3114.

Craufurd, D., Harris, R. (1986) Ethics of predictive testing for Huntington's chorea: the need for more information. *Br. Med. J.* 293:249–251.

Fahy, M., Bloch, M., Robbins, C., Turnell, R.W., Hayden, M.R. (1989) Different options for prenatal testing for Huntington disease using DNA probes. *J. Med. Genet.* 26:353–357.

Gusella, J.F., Wexler, N.S., Conneally, M.P., Naylor, S.L., Anderson, M.A., Tanzi, R.E., Watkins, P.C., et al. (1983) A polymorphic DNA marker genetically linked to Huntington disease. *Nature* 306:234–239.

Hague v. Williams. 37 NJ 328, 181A Zd 345 (1962).

Hayden, M.R. (1981) *Huntington's Chorea.* Springer, Berlin and New York

Hayden, M.R., Hewitt, J., Kastelein, J.J.P., Wilson, R.S., Hilbert, C., Langlois, S., Fox, S., et al. (1987) First trimester prenatal diagnosis for Huntington's disease with DNA probes. *Lancet* 306:234–239.

Hayden, M.R., Robbins, C., Allard, D., Haines, J., Fox, S., Wasmuth, J., Fahy, M., et al. (1988) Improved predictive testing for Huntington disease by using three linked DNA markers. *Am. J. Hum. Genet.* 43:689–694.

Lamport, A.T. (1987) Presymptomatic testing for Huntington chorea: ethical and legal issues. *Am. J. Med. Genet.* 26:307–314.

McFall v. Shizp. No 78-17711 in *Equity* (July 26, 1978), cited in note 40 *Ohio St. L.J.* 409 (1979).

Meissen, G.J., Myers, R.H., Mastromauro, C.A. (1988) Predictive testing for Huntington's disease with use of linked DNA marker. *N. Engl. J. Med.* 318:535–542.

United States President's Commission for the Study of Ethical Problems in Medicine and Biomedical and Behavioral Research (1983) Screening and counseling for genetic conditions 44.

Quarrell, O.W., Tyler, A., Upodhaya, M., Meredith, A.L., Youngman, S., Harper, P.S. (1987) Exclusion testing for Huntington's disease in pregnancy with a closely linked DNA marker. *Lancet* 1:1281–1283.

Shaw, M.W. (1987) Testing for the Huntington gene: a right to know, a right not to know, or a duty to know. *Am. J. Med. Genet.* 26:243–246.

Truman v. Thomas. 27 Cal 3d 285, 611, p2d 902, 165 *Cal. Rptr.* 308 (1980).

Wasmuth, J.J., Hewitt, J., Smith, B., Allard, D., Haines, J., Starecky, D., Partlow, E., et al. (1988) A polymorphic DNA marker tightly linked to the Huntington disease gene. *Nature* 332:734–736.

Wexler, N. (1989) The oracle of DNA. In: Rowland, L., Wood, D.S., Schon, E.A., DiMauro, S. (eds). *Molecular Genetics of Diseases of Brain, Nerve and Muscle.* Oxford University Press, New York.

Whaley, W.L., Michiels, F., MacDonald, M.E., Romano, D., Zimmer, M., Smith, B., Leavitt, J., et al. (1988) Mapping of D4S98 /S114/S113 confines the Huntington's defect to a reduced physical region at the telomere of chromosome 4. *Nucleic Acids Res.* 16:11769–11780.

5

Applying Morality to the Nine Huntington Disease Cases: An Alternative Model for Genetic Counseling

In this chapter, we apply the account of morality presented in Chapter 2 to the nine cases that were discussed in the article reprinted as Chapter 4. Although our final recommendations concerning most of these cases are the same as those given in [the reprinted article], we think that we provide a much clearer explanation and justification of these recommendations. However, we do differ in some of our recommendations, and we show that the use of principles may actually be misleading in some cases. Most important, our systematic account of morality shows the cause of some of the dilemmas and enables us to provide a procedure for avoiding them.

Introduction

We now apply the account of morality given in Chapter 2 to the cases considered in the article reprinted as Chapter 4. We have chosen to adopt this somewhat unorthodox way of proceeding for several reasons. First, we think that the article is a very good one by a very distinguished group of people. The solutions they offer to the dilemmas they present are sensible and will, for the most part, commend themselves to most people acquainted with the issues. Applying our account of morality to these cases will enable the reader to compare the usefulness of two different methods of dealing with relevant moral dilemmas, so as to assess their usefulness. Because we will be dealing with cases selected by those applying the alternative method, the case selection is not biased in our favor. Finally, we hold that there is almost universal agreement on the moral judgments that impartial rational persons will make on most cases when they agree on the facts of the case; that the answers that we offer are so similar to those offered by the authors of the article confirms this position.

What we want to demonstrate in this chapter is that the account of morality that we have presented in Chapter 2 provides a better understanding of the judgments arrived at than does the alternative method. We think that this is important because it allows one to teach and train people to deal with these dilemmas. Further, it makes it much more likely that those who have been so trained will be better able to deal with new dilemmas. Even more importantly, we shall try to demonstrate that our account of morality may make it possible to formulate public policies that will eliminate some of these moral dilemmas. As we mentioned earlier, we are concerned that the novelty of some of the problems that arise from the new genetics may distort one's moral intuitions. A clear account of morality may help prevent that distortion and enable one to see the similarity between the new dilemmas and the dilemmas that we have been confronting throughout human history.

We shall repeat each case before we begin our comment on it, so as to make it easier to refer to it in our discussion. Because our primary point is to show the value of using the account of morality that we have presented, we will be providing not only a short answer, similar in length to that presented in the reprinted article, but also a more extended discussion of the solutions that we propose than was done in the article. We realize that this alone provides an unfair comparison. However, we are convinced that, even were the explanations in the article expanded, they would not provide the kind of systematic account that could be usefully applied by other people to new dilemmas. It is because we believe that our account can be usefully applied to new dilemmas that we present our analysis of these cases even though we know that, the Huntington disease gene (*HD*) having now been discovered, the precise dilemmas presented here are no longer ones that counseling centers will face.

Case 1: Issues of Confidentiality

Two siblings requested predictive testing. The probands' (patients') mother, her maternal cousin, and the maternal grandmother were affected with Huntington disease. DNA analysis (see Appendix, Figure 4) revealed that proband III-2 is at high risk (97%) of having inherited the gene for Huntington disease, wherease proband III-1 has a low risk (approximately 3%).

In this family, the brother and sister were seen together in the early stages of the program; so each clearly was aware that the other was having the test. Proband III-2 later elected not to inform his sister of his own increased risk result and instead told her that there had been some delays in the laboratory and that his results were not yet available. The sister, proband III-1, at the time she received her own results, expressed regret that her brother had not yet been able to have any news from the test and enquired when he might receive his results. The counseling team was aware that the brother had been given results and that he obviously had not informed his sister of his results.

Ethical Analysis

There seems to be no problem here. No moral rule requires the counseling team to answer the sister's question; they can say that they cannot discuss his case with anyone else. In this case, not telling is not a violation of the rule against deceiving. Even if one thought that informing her that her brother had received his results would discourage deception and hence be following a moral ideal, that is not sufficient to justify breaking the promise of confidentiality to the brother.

Extended Analysis

In coming in for testing, patients are assured that their test results will remain confidential and that no one will be informed of their test results without their permission. In this case, a brother and sister are both being tested; the results of testing are that the brother almost certainly (97%) has the *HD* gene and the sister almost certainly (97%) does not have it. Although both the brother and the sister have been told of their own results, the brother does not want his sister to know of his results and, in fact, lies to her. This lie involves the testing facility, because he claims that he has not been told of his result owing to delays on the part of the facility. An extremely plausible explanation of why he told this kind of lie instead of simply trying to reassure his sister by telling her that he, like she, almost certainly did not have the *HD* gene is that he felt that he would have been unable to feign the kind of relief that she would have expected.

In the absence of this lie involving the testing facility, there seems to be nothing that even remotely warrants breaking the moral rule against breaking the promise of confidentiality or the duty of the facility to maintain privacy. For purposes of ethical analysis, it makes no difference whether one is morally required to maintain confidentiality as the result of an explicit promise or because it is simply taken as a duty of the testing facility, as with most medical interactions, to maintain privacy except for clearly specified exceptions. Apart from the brother's lie about the testing facility, there are no morally relevant considerations that provide even the slightest grounds for advocating that violating either moral rule be publicly allowed—for example, there are no significant evils to be prevented or avoided by breaking the promise concerning, or the duty regarding, privacy.

However, because the brother lied about the testing facility, it may seem as if the counselors were collaborating in his deception of his sister by not informing her that her brother had indeed received his test results. Suppose that she asked when her brother was going to get his test results. How could one answer this question without being deceptive? It may seem as if one would be required to lie to her, and this violation of the rule against deception itself needs justification. However, this particular problem can be easily solved by taking advantage of the explicit policy concerning privacy and replying to all questions about her brother with the statement that all aspects of a patient's interaction with the facility are confidential, including even whether or not he has received his tests results.

It could be argued that, if the testers tell the sister that they had completed the tests for her brother, they would not be violating their promise to maintain confidentiality. It could be claimed that not only would they not be revealing the outcome of the tests, thus not violating the promise (duty) of confidentiality, but they would also be acting on the moral ideal of discouraging deception, because their telling makes it more likely that the sister will become aware of her brother's lie. However, especially in these circumstances, this would involve far too narrow an interpretation of the promise (or duty) to maintain confidentiality. Telling the sister that the test had been performed makes it almost inevitable that her brother cannot keep her from knowing the results of that test. Although the testers can easily avoid deceiving the sister by referring to the policy of the facility to maintain confidentiality, this method of avoiding deception does nothing to discourage the deception of the brother. The moral ideals encourage, though they do not require, that one prevent moral rule violations, including deception, unless this involves an unjustified violation of a moral rule.

The question that remains is whether the testers are justified in breaking their promise (duty) of privacy in order to follow the moral ideal of preventing deception. One is justified in breaking a promise (or any other moral rule) if and only if this kind of violation can be publicly allowed. This means that an impartial rational person can favor this violation even if everyone knows that this kind of violation is allowed. Publicly allowing the violation of their promise (duty) to maintain privacy in order to prevent deception to maintain that privacy would run completely counter to the point of the promise (or duty) to maintain privacy. One strong reason for promising confidentiality is the belief that some people will not come to be tested unless they are guaranteed that their test results will not be revealed to anyone else without their consent. Imagine trying to defend publicly allowing violating a promise (or duty) to maintain privacy by claiming that one is attempting to prevent deception by that person in order to maintain his privacy. It should have been apparent to those making the promise (assuming the duty) to maintain privacy that the proband might think deception necessary to maintain his privacy. Thus, if this kind of violation were publicly allowed, promises (or duties) to maintain privacy would lose so much force that those people who desire privacy would be discouraged from coming in for testing. Hence, as long as the testing group is not required to participate in that deception, deception by the proband cannot serve as a justification for violating the promise to maintain privacy; in this kind of situation, no impartial rational person could publicly allow such a violation.

However, it might be worthwhile to talk to the brother about his deception. While continuing to assure him that you will be extremely careful not to reveal anything, counseling should be offered. He should be offered assistance in telling his sister of his results. Although it is not required that testers encourage patients to be honest with others, it is, as indicated earlier, following a moral ideal to discourage deception. It is clearly not required to be neutral on this matter. It is morally required not to harass or pressure the brother to tell when he is not prepared to do so, because pressure and harassment violate the moral rule against causing pain, and in this case there is not an adequate reason for such a violation.

In order to make clear how the same kind of reasoning can be applied to similar cases, consider the following case. Suppose that the brother lies, not to his sister, but to his employer. The employer now asks the testing facility to verify the claim that the results have been delayed. In this case, the answer should be the same as before. The facility should take advantage of the explicit policy concerning privacy (confidentiality) and reply to all questions about the brother with the statement that all aspects of a patient's interaction with the facility are kept confidential, including whether he is even a patient of the facility, whether he has been tested, or if the test results are available.

What if the brother lies to his employer by claiming that he has been told that he almost certainly does not have the Huntington gene? Again, it seems as if the reply should be the same. However, just as before, the testing facility should not aid in the deception but stay as completely neutral as possible. It should not, for example, back up the brother's claim that he has received good news about the possibility of getting Huntington disease. It should not violate the promise (duty) to maintain privacy; nor should the testers themselves engage in deception. There is no morally relevant feature in this situation that would enable an impartial rational person to publicly allow either violating a promise to maintain confidentiality or violating the rule against engaging in deception.

In both of these cases, the moral system makes it clear that the proper response by the testing facility is to abide by whatever moral rule—namely, "Keep your promise," or "Do your duty"—requires it to maintain the brother's privacy. The moral system also makes clear that in these cases it is not required to take part in the deception. Confidentiality can be maintained by referring to the promise (duty) to maintain confidentiality without any deception. The testers are not justified in engaging in deception themselves. Finally, they are encouraged to follow the moral ideal of discouraging deception but are not allowed to do so if this involves an unjustified violation of a moral rule.

Applying the moral system in this explicit fashion makes clear that one may be faced with a more difficult case. Suppose the brother lies in a way that requires the testing facility to engage in deception, or else to violate their promise (duty) to maintain confidentiality. Suppose the brother tells his sister that he has authorized the facility to tell her his results, although he has told the testers that he does not want her to know that he has a high probability of having the *HD* gene. What is to be done now? They seem to be forced either to deceive or to break a promise (violate a duty). Either choice they make seems to have serious problems, because both choices compromise the integrity of the facility. When faced with this kind of situation, in which equally informed, impartial, rational persons may make different choices, one is said to be in a moral dilemma. Sometimes one cannot avoid a moral dilemma and is forced to make a choice, but clearly it is far better to avoid the dilemma if one can do so without sacrificing anything of moral significance.

In this case, the moral dilemma can be avoided by being more careful in how one formulates the promise (duty) to maintain confidentiality. The facility should have an explicit and public policy that, although it will not reveal the deception of a proband, it will also not itself engage in deception. The facility should promise only that it will not reveal, without the consent of the proband, anything that would compromise the confidentiality of the proband, but it will not participate in the deception of others. This clear public statement of policy concerning confidentiality eliminates the dilemma without sacrificing anything of moral significance. Indeed, it enhances the credibility of the facility, because it shows not only that the testers are aware of the moral dilemmas that might arise and have taken steps to avoid them, but also that they are providing the kind of information required if one is to have genuine valid consent to undergo testing.

An explicit public statement of the policy on confidentiality is clearly a useful method for avoiding moral dilemmas. If it is known that confidentiality will be scrupulously maintained but that the testing facility will not assist anyone in his attempt to deceive others, then one has eliminated many possible cases of moral conflict. Explaining this policy to prospective probands is an excellent way of informing them of the problems they may have to face and of what they can expect from the testing facility in dealing with those problems.

This policy does raise one significant question. If it is known that the proband can authorize the testing facility to release information to others, then his failure to officially do so may lead many to conclude that he is not providing them with the correct information. We do not see that there is any alternative to this policy, because refusing to give out any information even with official authorization seems to have even more serious problems. One advantage of letting people know of the limits that are being placed on their ability to maintain privacy by deceiving is that they may be much more reluctant to deceive; one disadvantage is that they may be more reluctant to be tested. It is an empirical matter whether the advantages of having any particular explicit public policy outweigh the disadvantages or vice versa. However, it seems clear that, once one has come to realize the possibility of these dilemmas, it would be a violation of the duty to provide informed consent not to tell the proband of the possibility that his privacy may be overruled. Further, if one has a policy, it is clear that prospective probands should be informed of it. Thus there is no doubt that one should have some explicit public policy concerning confidentiality and its limitations, and the only question is about its details.

One advantage of using the moral system is that it helps one to spot possible moral dilemmas before they arise and enables one to formulate a policy that avoids them. Many dilemmas arise from a promise (duty) to maintain confidentiality that is stated without consideration of the proper limitations to be put on that promise. Recognizing that, contrary to the claims of the authors of the article reprinted as Chapter 4, confidentiality is not a basic principle or moral rule, but that it gets its moral force as the result of a promise or a duty enables one to formulate that promise (duty) so that one avoids the moral dilemma that often arises when two moral rules conflict. Thus the moral system emphasizes the moral importance of having a well-crafted public policy concerning confidentiality, one that clearly and explicitly states the limitations and exceptions to the promise (duty) to maintain confidentiality.

Case 2: Unconfirmed Diagnosis in a Family Member

The proband (II-1) presented for predictive testing and was given a risk of 80% of having inherited the *HD* gene. This was a partially informative result because of the availability of DNA from only a few family members, including one affected person. Thereafter, we learned from relatives that a sibling of the proband was probably showing early signs of Huntington disease. This was communicated to us independently on a few occasions by family members. However, this sibling had not consulted a physician for many years and had never been formally diagnosed as having Huntington disease. He had previously given blood—and consent—for the DNA analysis that was to be used to reconstruct haplotypes of the affected parent. Even though he knew he could participate in the program, he did not want any modification of his own risk relative to *HD* but gave blood, which might increase the informativeness for other relatives. The DNA analysis revealed that the proband and his sibling had inherited different chromosomes 4 from their affected parent.

If the sibling of the proband indeed had Huntington disease, then the proband would have a very low risk of developing the disease, contrary to what he had been told. In order to maximize the predictive-testing information for our proband, suggestions were made by family members to encourage the sibling to have medical intervention.

Ethical Analysis

There is no moral rule requiring the counseling center to ask the sibling to come in for a clinical assessment; however, it does seem to be following a moral ideal to do so. The important question is whether it is possible to ask the sibling to come in without causing him any harm. If there is any significant risk that asking him to come in will cause him substantial suffering, it should not be done, because it is not morally justified to break the moral rule that prohibits causing mental suffering to one person in order to relieve the anxiety of another.

Extended Analysis

One would not violate any moral rule by simply asking the sibling to come in for clinical assessment. Even though he does not want any information himself, he has been cooperative in providing blood so as to help other family members find out if they have the *HD* gene. However, because it seems as if the sibling would gain information that he does not want to have from the clinical assessment, it is very important to tell him that he would probably learn from the clinical assessment whether or not he had Huntington disease. If he is indeed having early signs of Huntington disease, then it might be to his long-term benefit to learn it now so that he can plan accordingly, but that is for him to decide, and he must be told that coming in will probably result in his gaining that knowledge. If there is some reason to believe that the sibling would be significantly hurt by being encouraged to have a clinical assessment, it is not justified to encourage him to do so, even though the new knowledge gained would very significantly affect the information concerning the probability of the proband having the *HD* gene. However, if one does not think that he will suffer any significant harm by simply being asked to come in while being provided with the information that he may find out about his own Huntington disease status, then one is following a moral ideal to tell him about the proband's request. Conveying the request provides him with an opportunity to help the proband and gives him an occasion for changing his mind about finding out his own diagnosis.

The sibling could be told, correctly, that having a clinical assessment would aid in determining the probability of the proband's having the *HD* gene. However, neither he nor anyone else would be told whether his having Huntington disease counts for or against the proband's having the *HD* gene. If it is possible for him to have a clinical assessment without finding out the results, both the sibling's desire not to know and the proband's desire to know might be satisfied. This would provide everyone with all of the information they want to know without violating any confidentiality considerations. No moral rules are broken, the promise of confidentiality is maintained, and no one is hurt; hence nothing needs to be justified. In addition, one is able to satisfy the rational request of the proband. This depends on the clinical assessment not revealing the results to the sibling, and this may not be a realistic expectation. Or the sibling may change his mind about wanting to know if he has Huntington disease, which is somewhat more realistic, but this requires that the sibling be informed that he may gain such knowledge prior to coming in.

It would not be morally allowed to pressure or harass the sibling, because this would be breaking the moral rule against causing pain (mental suffering), and the only benefit would be to relieve mental suffering of approximately the same magnitude on the part of the proband. When the evil being prevented is not unquestionably significantly greater than the evil caused, it is not morally allowed to break a moral rule. Given the natural biases in favor of oneself, one's friends, and one's family, publicly allowing violations, when the evil prevented by violating a moral rule is not unquestionably significantly greater than the evil caused by the violation, would lead to an increase in the overall amount of evil being suffered. Thus no impartial rational person would publicly allow those kinds of violations.

Another easily answered question is whether the proband should be told that he and his sibling have different chromosomes 4, and hence, if one has HD, the other does not, and vice versa. If this information were to be given to the proband, one would be providing him with information about his sibling, and this violates the kind of confidentiality that has been promised to those who supplied blood to aid in the testing of other family members. This information is very likely to lead the proband to pressure the sibling to have a clinical assessment so that he may gain more complete information about himself. Here again, the violation of a promise of confidentiality is not justified in order to reduce the proband's level of uncertainty and consequent anxiety, especially when violating that promise has a high likelihood of the sibling being subjected to very unpleasant pressure. Therefore, the proband should not be told that the siblings have different chromosomes 4, nor should any pressure be put on the sibling to undergo clinical assessment.

Suppose the sibling continues to not want to know whether he has Huntington disease. It may not be possible for the sibling to have a clinical assessment and not find out whether he now has Huntington disease. Here, as in so many other cases, what is crucial in deciding what alternative to choose is knowing the facts of the particular situation. How likely is it that the sibling can come in for a clinical assessment and not find out whether he has Huntington disease? How much would he suffer simply by being asked to come in for such an assessment? If the sibling would suffer from any attempts to have him receive clinical assessment, then one can only attempt to help the proband live with the uncertainty, telling him that he will be kept informed as new information becomes available; for example, if it becomes clear that the sibling does have Huntington disease, the proband could be told that he does not have the *HD* gene. The moral rules do not require that the counselors undertake all of these extra efforts to provide the proband with more information but, given that they can do so without breaking any moral rules, including not causing any additional suffering to the possibly affected sibling, it is following a moral ideal to do so and so is a morally good way to act.

Case 3: The Problem of Too Much Information

Proband II-1 has requested prenatal exclusion testing and would prefer not to have any modification of his or her own risk for Huntington disease. This type of testing has been described elsewhere (Chapter 4, References: Quarrell et al. 1987; Fahy et al. 1989) and could certainly be done in this family. DNA analysis would include the parents (II-1 and II-2) of the fetus and the affected grandparent (I-1).

Concurrently and without the knowledge of proband II-1, a relative (II-3) requested predictive testing. In order to undertake the analysis, we obtained blood from additional crucial relatives in the family. This resulted in information that proband II-1 is at low risk of developing Huntington disease, and prenatal testing could be considered unwarranted. However, proband II-1 has requested prenatal exclusion testing; the express wish was to receive no alternation in his or her own risk for developing Huntington disease. This would require that we ignore the knowledge gained serendipitously and that we provide only the test that the candidate requested, resulting in prenatal testing on a fetus that is known to be at low risk.

It could be argued that it would be in the best interest of the candidate to disclose that he or she has a very low risk of developing Huntington disease and that further testing is unwarranted. This approach may significantly reduce the proband's anxiety and stress levels, and it also avoids the unnecessary risk and expense of prenatal testing.

However, the situation would become increasingly complex if other family members requested prenatal exclusion testing. Siblings of proband II-1 and others might become aware that we have a policy of disclosing serendipitous discovered "low risk" results to prevent an unnecessary prenatal procedure. Any siblings who are not told that the prenatal test is unwarranted might infer that their own risk of having inherited the *HD* gene is high.

Ethical Analysis

This situation presents a genuine moral dilemma. A promise of complete confidentiality has been made, and it now turns out that keeping that promise requires one to cause a medically unnecessary risk of an unwanted spontaneous abortion. Thus one must violate a moral rule: either break a promise or cause some harm. Here one must be clear about the facts, particularly how great the risk of harm is, and then ask whether, given this risk, one would publicly allow the breaking of the promise or not. We think that impartial rational persons might give different answers to these questions, which is why we consider it a genuine dilemma. However, once one realizes that the dilemma is caused by making a more sweeping promise than is warranted, one can avoid the dilemma by explicitly formulating a policy that states that confidentiality will not be maintained if it requires causing an unnecessary risk of harm or prohibits preventing significant harms from occurring.

Extended Analysis

This case, like Case 1, concerns the limits of confidentiality or privacy. In this case, keeping confidentiality requires performing an unnecessary prenatal testing on a fetus, which has one chance in three hundred of causing an unwanted spontaneous abortion. Although this is not a high-risk procedure, in this case there is absolutely no medical reason for performing the prenatal test. It is already known that the fetus is at low risk for the *HD* gene, and the test cannot reveal any new information on this matter. The only reason for performing the test is to assure that future clients who want exclusionary testing will be able to come in for it without thinking, incorrectly, that they are able to infer their own *HD* status simply from the fact that the test is being performed. (This is an incorrect inference because the test may be performed because one does not have sufficient information to make the test unnecessary.)

This case is far more troubling than Case 1 because, in this case, there is a real risk of harm being caused in order to keep one's promise of maintaining confidentiality. This harm is not only to the fetus, which may not survive because of a spontaneous abortion, but also to its parents, who clearly do want a child who is at low risk for inheriting the *HD* gene. It is certainly true that, if one has promised exclusionary testing to the parents and others, then one has a moral dilemma. Either one causes an unnecessary risk or one breaks a promise. It is not completely clear how one decides this issue. We cannot say with certainty that all impartial rational persons would decide in one way rather than in another. However, in this case even more than in Case 1, the important point is to understand what caused the dilemma and to see if one can avoid it.

Again, just as in Case 1, the dilemma can be avoided by having an explicit public policy that makes clear the limitations on confidentiality or exclusionary testing. As the description of the case makes clear, the importance of a public policy is not appreciated by the authors of the original paper. This is evident from their statement, "Siblings of person II-1 and others might become aware that we have a policy of disclosing serendipitously discovered low risk results to prevent an unnecessary prenatal procedure." But, if the testing facility states explicitly and publicly that it will perform no unnecessary prenatal tests even if that results in people being able to infer something about their own *HD* status, then they can discuss this issue with prospective clients, allowing them to make an informed choice. Parents who request prenatal exclusion testing already are at risk for the *HD* gene and simply want to know if the fetus is at the same risk. The facility's concern that "Any siblings who are not told that the prenatal test is unwarranted might infer that their own risk of having inherited the gene for HD is high" can be dealt with openly. And, because it is unlikely that the testing facility will often have "too much information," the facility can make clear that this unpleasant inference is usually mistaken.

It seems to us that even with a risk of only one in three hundred of a spontaneous abortion, a public policy of not performing unnecessary prenatal tests would be preferred by more people than the public policy of maintaining confidentiality at all costs. It seems to us that preference for the latter policy does not take into account that all policies of the testing facility must be public policies. Informed-consent considerations require that the clients know what the policy is and what consequences it has for them. They should be informed that the latter policy subjects them to a risk of an unnecessary spontaneous abortion. It may be that some people are so determined not to know whether or not they have the *HD* gene, that they will take this risk, but it also seems that many would not. Because one is unsure of the preference of those concerned, families should be given a choice of whether they want exclusionary testing so much that they are willing to undergo this risk or whether they would prefer to give up confidentiality in order to avoid such an unnecessary and unwanted risk.

We do not, at this point and without much more empirical investigation, claim that any one of the following three public policies is preferable: (1) no unnecessary tests even if confidentiality is breached, (2) no breaching of confidentiality even if it requires performing an unnecessary test, (3) families will be told of the possibility of this dilemma and will choose which alternative they want to apply to them. There are arguments for all three. The first lets all prospective clients know that no unnecessary prenatal tests will be performed but gives them slightly less confidentiality protection. The second provides maximum protection for confidentiality, but at the risk of unnecessary prenatal tests. The third seems to be the best public policy, allowing the prospective clients to choose whether or not they want limits on their confidentiality, but it is not clear if it is a workable policy. In any event, having any of these public policies eliminates the moral dilemma. The first does so by not promising confidentiality when the prenatal test is unnecessary; the second does so by having patient consent for undergoing the unnecessary risk; and the third does so by having the prospective client make an informed choice between policy one and policy two as the policy that applies to them.

It may be that there are considerations that we have not considered that will affect which public policy one adopts—for example, they each might result in different rates of couples coming in for testing. But it is certainly better to have a public policy that lays out the consequences of maintaining confidentiality than not to have any public policy at all. In the latter situation, which seems to be the situation that the testing facility is in regarding Case 3, the testing facility is making a decision about a matter that the clients have not considered at all, let alone made an informed decision about. Imagine the worst-case scenario, the prenatal test is performed, there is a spontaneous abortion, and the couple discovers that the prenatal test was unnecessary. Imagine their reaction. At least if they knew of this risk and chose it, they could not hold the testing facility responsible for the unwanted abortion.

Clearly, what the consequences of adopting these different public policies would be is an empirical matter, one that should be subject to genuine empirical testing. How many, if any, people would be discouraged from coming in for prenatal testing if the first policy of performing no unnecessary prenatal tests were adopted? If the second policy is adopted, how severe would the reaction be of those couples who have spontaneous abortions and who do not know whether the prenatal test was unnecessary? If the third policy is adopted, do people feel that they are being subjected to a significant burden when they have to choose between the two policies, and what percentage choose which policy to apply to them?

What is quite clear is that it is morally required to have some public policy so that prospective clients can make an informed decision on this matter. But, although it may seem as if the third policy, letting the prospective client determine which policy applies to them, is the preferable policy in this case, it is clear that this kind of policy is not the best policy for all kinds of genetic tests. Prenatal testing for *HD* is only one kind of genetic test. The information gained from the test will be used by the couple to determine whether or not to have an abortion. No one other than the couple and their fetus is involved. But sometimes there will be other adults who can be affected; so the situation resembles, in some ways, testing for HIV.

This can be seen more clearly when one considers a genetic disease that differs from Huntington disease in a morally significant way. Huntington disease is an adult onset disease with no treatment and no means of prevention. The primary benefit that comes from knowing whether or not one has the *HD* gene is psychological; one no longer has the uncertainty and the anxiety that goes with such uncertainty and can make more definite plans for the future. If one finds out that one does not have the *HD* gene, then, except for some problems with survivor guilt, the benefits seem to outweigh the costs. Even if one finds out that one does have the *HD* gene, living with the certainty that one will develop Huntington disease seems to cause less anxiety and depression than does living with uncertainty. Nonetheless, it does not seem irrational not to want to know whether or not one has the *HD* gene.

Matters are very different with some other adult-onset diseases—for example, some forms of early breast cancer. Knowledge that one has the gene allows one to take preventive measures that can lengthen useful life by 20 to 30 years or more. Determining with high probability whether one has the breast cancer (*BC*) gene requires the same testing of family members that was necessary in testing for the *HD* gene. But the fact that knowing that one has the *BC* gene, unlike knowing that one has the *HD* gene, enables the person who knows it to take appropriate preventive measures is a sufficiently important difference that it should lead one to adopt a different public policy concerning confidentiality. There are many other cases like this type of early breast cancer—for example, familial adenomatous polyposis, which usually leads to colon cancer.

There is now in France a related genetic issue concerning confidentiality. This involves computer files of genetic information such that one can discover who is at risk for glaucoma and provide them with medical treatment that will completely prevent blindness from resulting. By giving privacy primacy over all other concerns, one loses the opportunity of informing affected persons of the measures that they could take to prevent blindness. It is not clear that if people knew that this would be one of the results of complete confidentiality that they would choose it over a policy of a more limited protection of confidentiality. In general, it seems as if, when the choice is between preventing serious harms (e.g., premature death or significant permanent disability) and complete protection of confidentiality, the former would usually be chosen. However, this is not certain. Some may feel that, once any breach in confidentiality has been made, there will be no way to stop confidentiality from being completely eroded, and so it is worth suffering a significant amount of harm in order to protect that confidentiality.

We think that this "slippery slope" argument has great force when one claims to have a policy of strict confidentiality and then breaks it in order to prevent a serious harm. That is why we think that Case 3, as originally described, does present a genuine moral dilemma. There are strong reasons for not violating a promise of confidentiality and strong reasons for not undergoing unnecessary risks. We do not think this slippery slope argument has much force when one is talking about establishing an explicit public policy. If the public policy is to protect confidentiality except in cases in which giving someone information will enable that person to prevent serious harms (e.g., premature death or significant permanent disability), we do not see why that is any more open to abuse than a policy of absolute confidentiality. Indeed, an argument can be made that the explicit public policy of limited confidentiality, by removing the strongest reasons for violating one's public policy, will be much less liable to erosion.

It is important to point out that one reason for having a policy of complete confidentiality in testing for the *HD* gene is that people who do not want others to know may not be willing to come in for testing, and hence one will be unable to aid those who want to know whether they have the *HD* gene but want confidentiality. It is rational to want not to know whether or not one has the *HD* gene, because given one's temperament, the evils of knowing may outweigh the evils of uncertainty. (However, recent studies seem to show that for most people knowing whether or not they have the *HD* gene is psychologically more advantageous than living with uncertainty.) With the *BC* gene or the colon cancer gene, however, it is irrational not to want to know, for knowing will enable one to prevent or significantly postpone the serious evils that will result from having the gene and not knowing. It will also make it unnecessary for one to subject oneself to the serious, though less serious, evils of acting to prevent or significantly postpone the serious evils that will result from having the gene, when one does not have it. Thus it seems that any rational person would want to know whether or not she has the *BC* gene.

Given this fact, a different policy concerning confidentiality is appropriate. It should be an explicit policy that any person who comes in for testing should be told that any female siblings will be informed that they should be tested if the proband has the *BC* gene, or if there is a significant probability that her mother had the *BC* gene. The siblings need not be told whether the proband has the *BC* gene, or even that the proband has come in for testing, but they will know that the proband is at risk for the *BC* gene. The reason for adopting this policy is that it will not discourage many people from coming in for testing, because, if a women suspected that she might have the *BC* gene, it would be irrational not to be tested. It will also allow the testing facility to avoid the moral dilemma that might arise in the unusual case in which the proband does not want any of her female siblings informed.

Because there is one chance in two of helping someone to prevent or significantly postpone the serious evils that will result from not knowing that one has the *BC* gene, it is impossible not to be strongly tempted to inform the female siblings of the proband of their situation even if one has promised complete confidentiality. Having the explicit public policy just outlined avoids this moral dilemma by making it clear from the start that, in coming in to be tested, one waives confidentiality to the extent necessary to help prevent preventable death and disability. Apart from this explicit exclusion, the standard policy of confidentiality would be maintained.

A public policy of strict confidentiality has many advantages for a proband, and there is no point in limiting confidentiality except when it leads to disadvantages—for example, requiring one to engage in deception or to take the unnecessary risk of prenatal testing—or to forego significant advantages—for example, preventing or postponing very significant evils, such as premature death. In each of these cases, one must weigh the advantages of a public policy of strict confidentiality against the advantages of a public policy of confidentiality with limitations. In adopting a public policy on confidentiality, one must consider that genetic disorders may differ from each other in their morally relevant features—for example, whether or not knowledge of one's having the gene will enable one to prevent or postpone very significant evils. An advantage of a confidentiality policy with explicit limitations is that it helps one avoid the moral dilemmas that may arise in the absence of those explicit limitations. Another advantage is that it makes it much more likely that the probands will have a better understanding of the nonmedical risks of genetic testing and so be able to provide a more informed consent to such testing.

One can make use of moral dilemmas in order to discover where it might be useful to redesign the public policy of confidentiality by incorporating some additional explicit limitations. The moral system makes clear that confidentiality is not, as maintained by some, a basic principle, but that the degree to which one is required to maintain confidentiality depends entirely on the terms of the promise (duty) to maintain that confidentiality. Thus the primary benefit of using the moral system may not be its helpfulness in dealing with particular moral dilemmas after they have arisen, but in preventing them from arising in the first place.

Case 4: Reconstructing Haplotypes

In this family, the proband's father (I-2) dies at age 36 years, with no clinical signs of Huntington disease, and it is not known whether he had inherited the *HD* gene. There is a documented family history of Huntington disease in that the paternal grandmother and several other individuals were affected. In order to provide definitive information for the proband, the father's haplotypes could only be reconstructed using DNA from his children (persons II-2, II-3, II-4, and II-5). However, the proband (II-2) is the only individual in this sibship who has requested testing. If the siblings' DNA samples are analyzed for reconstruction of the father's haplotype, then information could be learned regarding their individual risks of having inherited the *HD* gene.

Ethical Analysis

There is no moral rule requiring one to ask the siblings to come in for a clinical assessment; however, it does seem to be following a moral ideal to do so. All that we know is that they have not requested testing; we do not know that they have refused it. It may be possible to ask the siblings to consent to be tested without causing them any harm, but asking them to contribute DNA samples must be accompanied by providing them with the information that they may come to know of their risk of having inherited the *HD* gene by those who are not part of the counseling facility.

Asking them to contribute DNA samples without providing them with this information should not be done, because they would not then have the information necessary for informed consent. Withholding such information in order to get them to contribute DNA samples counts as a violation of the rule against deceiving. It is not justified to deceive in order to satisfy the rational desire of the proband to know when this will result in thwarting the rational desires of others not to know.

Extended Analysis

As stated in our extended analysis of Case 1, an explicit statement of policy on confidentiality is clearly a useful method for avoiding moral dilemmas. In this case, the siblings could be approached in a respectful manner (without harassment or pressure), presented with a variety of options, and offered counseling and relevant information to help them decide. These options would include not donating blood; donating but not receiving any further information or counseling about their own or anyone else's risk status, but allowing the information obtained to be available to the proband or other interested family members; donating blood and receiving their own risk information, but not receiving information about other family members; or donating and receiving information about both their own risk and that of other family members who have consented to the disclosure of risk information. They must also be told that they may find out about a change in their chances of having *HD* regardless of whether or not they choose to be told.

The testing facility cannot guarantee that a sibling wishing not to know his own risk status will not obtain such information through the disclosures of other family members about *their* status. For example, if the proband turns out to be at high risk for Huntington disease (not the case in this instance) and chooses to disclose this information to his other sibs, they would then know of their own increased risk, even though the facility itself has honored their requests not to know of any change in their own risk status. Presumably all the sibs knew that their paternal grandmother had the disease, and thus that their risk of developing it was 25% (even if the involved facility does not provide this basic sort of information, it is readily obtainable from other sources). If the proband discloses his high risk to his sibs, they can easily find out that their own odds have just increased from 25% to 50%. This might tempt them to want to know more precise information about their risk whereas, if the proband had said nothing, they might have stuck to their initial wish not to know. There are many similar possibilities and complications that might arise: people's wishes may change after they have given blood, a family member may lie about his results and wish to involve the facility in the lie (Case 1), and so forth.

It seems that a clear statement of policy on confidentiality, warnings about possible problems that can arise despite the best precautions, and a discussion of the general arguments for and against knowing one's risk for Huntington disease, coupled with an offer of counseling and ongoing support, are appropriate elements of the informed-consent process when obtaining blood for genetic analysis. Such a policy might state at a minimum: that, unless necessary to prevent serious harm, all interactions between the facility and an individual are confidential, but the facility will not aid in the deceptions promulgated by individuals about their test results; that the individual wishing not to know of any change in his status may learn of such change by virtue of disclosure by another family member, despite the honoring of confidentiality on the part of the testing facility; and that the possibility exists of discovering an unsuspected instance of nonpaternity through DNA testing (Case 9). It is possible that such a formidable list of warnings might discourage some people from being tested, but we do not believe this will happen if these risks, and the benefits of being tested, are presented in a sensitive fashion.

Case 5: Inability to Obtain Blood from a Crucial Relative

In this family, the affected mother (II-2) is deceased, and the only living relative with Huntington disease is her half-sibling (II-3), who is elderly and severely incapacitated with the disease and who has been deemed an incompetent person. Although the predictive test is unlikely to be highly informative in this family, it may be possible to provide at least some alteration in risk for those individuals requesting the predictive test. However, all efforts to obtain consent for blood withdrawal from the affected individual have failed, because of long-standing conflicts within the family and because of failure to get permission from the appointed guardian. Without a sample of DNA from this crucial relative, it is impossible to offer predictive testing for probands III-1 and III-2. The at-risk individuals have been investigating options open to them with regard to petitioning the courts to request the cooperation of the affected relative.

Ethical Analysis

As stated in our extended analysis of Case 2, when the evil being prevented is not unquestionably significantly greater than the evil caused, it is not morally allowed to break a moral rule. Only if one ranks drawing blood as a negligible harm, and the reduction of anxiety to the siblings as significant, might one consider it justified to try to persuade the guardian to allow blood to be drawn from the affected incompetent aunt. If drawing blood is viewed as a more serious harm, then it is doubtful that it is even justified for the guardian to allow the blood to be drawn.

Extended Analysis

The law gives the guardian the right to refuse the taking of blood from the incompetent person. The guardian does not seem to be acting to protect the interest of the incompetent aunt, but rather simply to frustrate the desires of the probands. Even though the guardian does not seem to be acting in a morally good way, this is not sufficient to justify the facility taking any action other than trying to persuade the guardian to act in a more compassionate way. And even this action should be undertaken only if it is clear that it would be at most a negligible harm to the aunt to have blood drawn. Contrary to what seems an assumption of this case, it may be that the guardian thinks that drawing blood will be a traumatic experience for the incompetent aunt with Huntington disease. If this is the case, then, unless one can prove that this is not true, the guardian should not allow blood to be drawn.

Even if the guardian is acting completely out of spite, there is not sufficient benefit to the probands to warrant the testing facility taking any action against the guardian. The motive of a person, though it reflects strongly on the moral character of the person, does not necessarily have any clear relation to the moral character of the act he is performing. Even if the guardian is acting out of spite, if the drawing of blood would risk causing significant harm to the aunt, then he is doing the morally right thing by refusing to allow it. One can do the right act for the wrong motives. Further, even if some impartial rational persons would hold that the guardian should allow blood to be drawn, if some would not (i.e., as long as impartial rational persons can disagree) it is not justified to try to force the guardian to change his mind.

Suppose that all impartial rational persons were convinced that the harm being prevented by the reduction in anxiety to the probands was so much greater than the harm caused to the aunt that it would be irrational for one not to favor a public policy of allowing the harm to be caused to the aunt. Unless there are some unknown facts, the supposition under consideration is obviously false. And only if this supposition were true might it be acceptable for the testing facility to put any pressure on the guardian to allow testing. It seems that, given the facts of the case, not only is the testing facility not morally required to do anything to pressure the guardian, it is not even clear that they are morally allowed to do anything. Luckily, the at-risk individuals have not asked the testing facility to do anything but are considering petitioning the courts themselves. The testing facility should not get involved.

Case 6: Problems of Terminating a Fetus at 50 Percent Risk of Developing a Late-Onset Illness

A woman in her early twenties requested prenatal diagnosis but did not wish to have any alteration in her knowledge of her own risk for Huntington disease. In this family, it was possible to clearly distinguish the maternal grandmother's marker from that of the affected maternal grandfather. The mother inherited marker *A* from her mother and marker *B* from her affected father. The fetus proved to have inherited marker *B*, implying that the fetus inherited a chromosome 4 from its affected grandfather. The likelihood that the fetus has inherited the gene for Huntington disease rose from 25% to close to 50%. The pregnancy was terminated at 13-week gestation (Chapter 4; References: Hayden et al. 1987).

Subsequently, there has been considerable ongoing contact with other family members, with much discussion centering on whether prenatal exclusion testing should have been offered. The unaffected parent of our proband has expressed concern for the future implications of having terminated a pregnancy at 50% risk for a late-onset illness. She has requested that such testing be avoided.

Ethical Analysis

The facility is being requested to have a public policy regarding testing that may result in abortion. That is generally a good idea; however, whatever policy the facility develops depends upon its view about abortion. There is an unresolvable disagreement about the moral acceptability of abortion (see Chapter 9); so it is impossible to recommend a public policy that will be acceptable to all testing facilities. However, given the legality of first-trimester abortion for no reason at all, a secular testing facility need not have a policy restricting testing during the first trimester. However, it might be worthwhile to provide counseling to those who plan to have an abortion solely because of the genetic condition of the fetus.

Extended Analysis

If the facility thinks that abortion, at least through the first trimester, is solely the choice of the mother, it is likely that the public policy will be not to place limits on testing for abortion. However, given that women who come for genetic testing would not choose to have abortions if the fetus is at no genetic risk at all, it is very important to make clear what the risks are and whether there are other tests that would provide a clearer account of the risks. Providing all relevant information in a way that encourages the woman to make her decision based on the best available information is the goal of all genetic counseling. It is difficult to do this while at the same time not attempting to persuade the woman to make one decision rather than another.

It is very important to distinguish between encouraging a woman to make a decision based on all of the available information and encouraging her to make one decision rather than another. Genetic counselors should encourage better decision making, but they should not encourage their clients to make the decision that the counselors would make. (For further discussion of the goals of genetic counseling, see Chapter 6.)

Testing facilities are not required to provide genetic tests to determine conditions that do not provide medical grounds for abortion (e.g., the gender of the fetus), no matter how early in pregnancy the test is done. When we are concerned with some later stage of pregnancy, at which abortion must be justified, the testing facility can limit testing to those genetic disorders that are regarded as providing justification. In general, the later the stage of pregnancy, the more serious the genetic risk must be in order to justify abortion. However, there is such variability in how serious this risk has to be at any particular stage of pregnancy that it is impossible to recommend any particular policy that should be adopted by all facilities.

It is, however, important that any policy adopted by a facility be one that all members of the facility agree on. This means that these policies should be adopted only after discussion with all members of the facility. If there is no consensus, then this itself is significant. Not only should the policy adopted be one that allows for these differences to be expressed, but it should also be a public policy—that is, it should be made known to all of those who come in for counseling that there are different views about testing held by different members of the testing facility. (For further discussion of the problems of abortion for genetic conditions, see Chapter 9.)

Case 7: Rights of Individual versus Rights of Society

We have had a request from a major air transportation company to do the predictive test on an employee without informing the latter—specifically, by obtaining a blood sample under false pretenses. This at-risk employee is a pilot, and the company wished to learn whether he is likely to have Huntington disease in the future, because it could influence his continued employment as a pilot. This pilot, at the time of the company's request, did not wish to have predictive testing because he felt that the information could be detrimental to his future career plans.

Ethical Analysis

The moral rule against deceiving prohibits obtaining a blood sample under false pretenses. This rule, like all others, can be justifiably violated only if one has a strong enough reason that this kind of violation could be publicly allowed. In this case, there seems to be no good reason for breaking the rule. Predictive testing would be less informative about the competence of the pilot than would regular tests of competency.

Extended Analysis

Although this case is presented as the rights of the individual versus the rights of society, as with many similarly labeled cases, this is extremely misleading. The testing facility is being asked to deceive one of its clients for the convenience of his employer. The employer has available to it many opportunities to test its employees for competence, none of which involve deception. There is absolutely no reason for the testing facility to violate the rule against deception in this case, and all talk of conflicting rights simply confuses the issue and makes it seem less straightforward than it actually is. This case, as presented, has an absolutely clear decision: deception is not allowed.

There might be a case, though it seems extremely unlikely to occur in real life, in which deception of a client might be necessary to prevent serious harm from occurring to very many other people. That it is hard to describe such a case without resorting to science-fiction scenarios shows how unlikely it is. However, suppose that there were a genetic condition that was fatal unless one underwent a treatment that was fatal to those not having the condition; and only by deceiving the client could one obtain a blood sample by which one could determine which people have this genetic condition. Even in this completely far fetched case, one would have to know far more about the time constraints that one was operating under and what efforts had been made to obtain valid consent from the client. In almost all actual cases, even if the claim about the amount of harm to be prevented by the deception were true, it would not be sufficient for impartial rational persons to publicly allow those kinds of violations.

Case 8: Predictive Testing for MZ Twins

One twin of a presumed monozygotic (MZ) twin pair at risk for Huntington disease requested predictive testing. Any results given to one twin would also apply to the co-twin, who did not desire to have any alteration in his or her knowledge of their risk to develop Huntington disease. To accede to the wishes of one twin is to deny the rights of the other twin.

Ethical Analysis

Is a monozygotic twin who requests any information about himself, which also applies to the other MZ twin, required to get the approval of the other MZ twin? Is an MZ twin who requests predictive testing required to get the approval of the other MZ twin? These two questions seem to require the same answer. Once one realizes this, it seems clear that the answer is no, he does not need the approval of the other MZ twin. All talk of rights seems out of place here; there is no right to have predictive testing and no right to have one's ignorance guaranteed. But to deny the test to the twin who wants to know simply because he has a twin who does not want to know seems to give one person an unjustified control over another.

Extended Analysis

Monozygotic twins seem to raise significant problems for genetic testing, especially if one talks about rights to autonomy and privacy. In the matter of genetic information for MZ twins, what one knows about oneself, one also knows about one's twin. However, if one remembers that each twin is an individual whose freedom to know cannot be curtailed by the wishes of anyone else, even his twin, then this case has a straightforward solution. The twin who comes in for predictive testing should be given that test. All reasonable efforts should be made to have the twins work out some arrangement that allows both of them to satisfy their wishes as far as possible. Further, a testing facility may have a public policy prohibiting providing predictive testing unless all who may be directly affected by the information give their approval. However, it is not clear why any facility would adopt that public policy rather than a policy that allows all people to decide for themselves whether or not they want to be tested.

Because we now know that people seem to fare better when they know whether they have the *HD* gene than when they do not, it might be worthwhile to try to persuade the twin who does not want to know to change his mind. But if that cannot be done, in the absence of a specific public policy, the nonconsenting twin does not have a veto over the twin who wants to know. In the absence of a policy, there is no justification for limiting the freedom of one twin in order to satisfy the desires of the other when both desires seem of equal importance. Even if, contrary to fact, the desire of the twin not to know, were slightly more important, that would still not be sufficient to limit the freedom of the twin who wants predictive testing.

Situations often arise in which one person's gain is another person's loss and vice versa. It may be useful to have as a goal harmoniously satisfying the desires of all persons involved; however, we all know that sometimes this is not possible. When this kind of situation arises, one must be careful not to violate a moral rule with regard to one person simply in order to prevent a harm to another, unless the harm to be prevented is undeniably significantly greater than the harm caused by the violation. In the absence of a policy, to deny testing to the twin who requests it is violating his freedom, and so the test should not be denied. Further, there seems to be no good reason for adopting such a policy. However, the experience of testing facilities may provide the kind of empirical information about the effects of providing and not providing predictive testing to one MZ twin when the other twin does not want to know, which could support such a policy (e.g., providing the test may result in far more family problems than not providing the test). In the absence of this kind of information, it seems that the facility should not allow one twin to veto the request of another for predictive testing. But this example shows that new facts can change the policy that a facility should adopt.

Case 9: Knowledge of Unsolicited Information

A man at risk for Huntington disease and his spouse have requested prenatal exclusion testing. He declined to have definitive testing for himself. DNA typing done prior to the prenatal test to identify which markers are informative revealed that the affected parent of the proband was not his biological father. That is, a person who requested prenatal testing because he thought he was at risk for Huntington disease was discovered not to be at risk for the disease. Should prenatal testing for *HD* be done in this family when it is known that the "at risk" parent is in fact not at risk for Huntington disease?

Ethical Analysis

This case, like that of Case 3, presents a genuine moral dilemma. A promise of confidentiality has been made to the parents of the proband, and it now seems that avoiding a medically unnecessary risk of an unwanted spontaneous abortion may require breaking that promise. This dilemma shows that the parents of the proband should have been informed at the time of testing that there were situations in which their privacy would be lost. However, it seems possible in this case, simply to tell the proband that sufficient information has been obtained that it is known that the fetus does not have the *HD* gene without doing any prenatal testing. It may not be necessary to tell him why, and indeed in this case one could refuse to give out further information.

Extended Analysis

Avoiding the risk of an unwanted spontaneous abortion should be one of the primary goals of any testing facility. Given this goal, public policies should be adopted that eliminate the temptation to perform medically unnecessary prenatal testing. Among these policies should be explicit limits to the privacy or confidentially promised to those who provide DNA samples. It should be part of the informed-consent procedure that people are told that no medically unnecessary prenatal testing will be done by the facility and that this may sometimes result in information that is normally kept confidential becoming known. Although the facility can promise not to tell why prenatal testing is not necessary, donors should be told that there might be a risk of discovery of nonpaternity, because this is the kind of information that most people are most concerned to keep private. If people are coming in to help parents discover their fetus's risks for Huntington disease, appropriate counseling may persuade them to take the slight risk that some information they would prefer to remain unknown may become known, when taking that risk will prevent the risk of an unwanted spontaneous abortion.

In the case under consideration, there seems to be no requirement to tell the proband anything more than that the fetus is not at risk for Huntington disease. There is no requirement that he be told how this was determined. The testing facility should have a public policy that they do not reveal any confidential information unless absolutely essential to avoid causing unnecessary harm or to prevent significant harm from occurring.

This part of the policy prevents the facility from being forced to explain why they do not need to do the prenatal testing. It also assures those who have been told that their privacy is not absolute that it will not be violated unless absolutely necessary to prevent harm or to avoid causing it.

In this case, three different moral rules seem to apply: (1) the rule prohibiting causing (even risk of) harm, (2) the rule prohibiting deceiving, and (3) the rule prohibiting breaking promises. In the present situation, where the appropriate public policy is not in place, it seems possible to inform the proband that prenatal testing is not necessary without deceiving him, and so not causing any risk of harm to the fetus. It is not clear whether or not one would be breaking any promise of confidentiality to the proband's parents. With the appropriate public policies in place, one can clearly deal with the case under discussion without breaking any of the three moral rules. Sometimes, unfortunately, the appropriate public policy is not in place and one must break at least one moral rule. Then one is limited to trying to design a public policy that will avoid this kind of dilemma in the future. Sometimes, even this is not possible, and then one must deal with the dilemma by seeing what violation one would be most willing to publicly allow. By providing a procedure that one can adopt in trying to determine which violation is preferable, as well as helping one to design public policies that prevent one having to violate a rule at all, we think that our account of morality has shown that it has real practical value.

6

Clarifying the Duties and Goals of Genetic Counselors: Implications for Nondirectiveness

In this chapter, we use the distinction between the moral rules and the moral ideals in order to distinguish between the duties and the goals of genetic counselors. We make clear that a primary goal in genetic counseling is that the client make a fully informed decision, one that is based upon her own values and full information about her options, and nothing else. We show that, although nondirectiveness is generally useful in achieving this goal, it is misleading to regard it as a requirement of genetic counseling; achieving this goal may sometimes require directive behavior by the counselor.

Moral Rules versus Moral Ideals

The purpose of this paper is to clarify some of the main duties and goals of genetic counselors. By doing so, we aim to cast light upon a common ethical problem that arises for counselors in their daily practices, the question of whether or not to be directive during the counseling session. In order to prepare the ground for these analyses of applied issues, it is necessary to first recall some ideas that were presented in Chapter 2, which provides a systematic account of moral reasoning. In that chapter, we pointed out that, although morality (the moral system) provides a guide to everyone's behavior, it is very important to distinguish between the different parts of that guide. For the purposes of this chapter, the most important distinction is that between moral *rules* and moral *ideals*. Everyone is required to obey the moral rules impartially all of the time; that is, it is often legitimate to punish people when they do not do so. The rules prohibit people from causing harm or doing those kinds of actions that increase the risks of harm being suffered. It should be recalled that all of the basic moral rules either are stated as prohibitions or can be stated in that way with no change in meaning—for example, Do not kill, Do not deceive, and Keep your promise (do not break your promise).

All rational persons agree that these kinds of actions (e.g., killing and deceiving) are so destructive to individuals and to society that morality prohibits committing them unless the circumstances are such that one would publicly allow anyone in those same circumstances to break the rule. Commission of these acts unless one could publicly allow such a violation is forbidden by every community of rational persons. The other basic moral rules prohibit causing pain, disability, or loss of freedom or pleasure; breaking a promise; cheating; breaking the law; and neglecting one's duty. Adherence to these moral rules is also required by all professions, unless one can advocate publicly allowing that kind of violation. It is unimaginable that there could be a profession that did not prohibit all of its members from breaking any of these rules (for example, inflicting pain or deceiving) unless the circumstances were such that they would favor everyone knowing that breaking the rule is allowed in those circumstances.

Morality also includes ideals that tell people to take positive action to prevent or relieve the harms suffered by others. Moral ideals differ in character from moral rules in that adherence to them is encouraged rather than required; that is, people are usually praised for following them and it is not legitimate to punish people for not doing so. Persons who are valued by their community, such as members of a profession, often strive to act on these ideals. One such moral ideal is to relieve pain. In most circumstances, a person who sees a fellow human being in pain but does nothing to offer comfort is regarded as callous. But, unless one has a duty to help in those circumstances, failing to provide aid is not violating a basic moral rule. In fact, acting on these ideals more than occasionally is fairly uncommon, whereas the moral rules are normally obeyed by everyone almost all of the time. One is required to obey the moral rules all of the time with regard to everyone. One is encouraged to follow the moral ideals at any time with regard to any person one chooses.

The Duties, Goals, and Ideals of a Profession

Just as certain basic moral rules and moral ideals govern the behavior of everyone in the community, professions formulate derivative duties and goals for governing the behavior of their members. These duties and goals are sometimes expressed in professional codes of ethics. These codes of ethics, like our common system of public morality, contain rules, usually called duties, that all members of the profession are *required to* obey and ideals, sometimes called goals, that all members of the profession are *encouraged to follow*. The goals of a profession are always derived from the moral ideals, but the professional duties are derived both from the basic moral rules and from the moral ideals, though in the latter case their scope is narrowed considerably. Legal and professional strictures are often levied against professionals who do not follow them. Thus a dentist who pulls teeth without administering anesthesia without a very strong reason not to give the anesthesia is likely to be sued in the civil courts and banned from practice by the profession. A dentist has a *duty* to provide such anesthesia. This duty, which is derived from the rule that prohibits causing pain, must be followed by all dentists. In some situations, a nurse has a duty to relieve the pain of her patients. This duty is derived from the general moral ideal to prevent pain but, when it becomes a duty, it's scope is drastically narrowed. No matter whether it is derived from a moral rule or a moral ideal, once it becomes a duty, the person is subject to penalties if she fails to act as required; for example, a nurse is derelict in her duty if she fails to administer the prescribed pain medication to her patient. In general, the particular specifications of the moral rules and ideals that are required of all members of a profession are the *duties* of that profession.

In addition to duties, all professions have *goals* that are derived from the ideals of the profession. They specify the harms which the professional is encouraged to prevent or relieve but that the profession has not made into a duty. For example, it is a goal of the veterinary profession to prevent and relieve the suffering of animals. Veterinarians are encouraged to engage in research and to educate the owners of animals as ways of achieving this goal. However, veterinarians are required to alleviate the suffering only of those animals that they are treating, not of all animals that they encounter. Given the fact that animal suffering is commonplace, this would be an impossible requirement. A veterinarian is, however, encouraged to act on the appropriate ideal. For example, a veterinarian is not required to treat a neighbor's dog for mange just because she is aware that the animal is suffering. However, it would be following an ideal for her to educate her neighbor about the dog's condition as a way of striving to achieve the goal of alleviating animal suffering.

Genetic counseling is a profession. Like all other professions, it has duties that are required of all members of the profession and ideals that encourage behavior that furthers the goals of the profession. Genetic counselors have a *duty* to provide accurate information about genetic tests. For example, a counselor who told her clients the outcome of an amniocentesis by flipping a coin rather than reading the test results could be legally prosecuted and expelled from the profession. This duty is derived from the basic moral rule prohibiting deception.

The counselor has a specific duty to avoid imposing her own values upon the client. A genetic counselor who tried to persuade a woman not to abort a fetus because of the counselor's religious beliefs would be violating a professional duty. This duty is derived from the moral rule prohibiting people from interfering with the free choices of others. Thus the code of ethics of the National Society of Genetic Counselors not only states that genetic counselors should give accurate and up-to-date information, but also states that genetic counselors respect the diverse values of their clients. Both are duties of this profession. For this reason, nondirectiveness is given great importance in the training and practice of this profession. It is the view of the profession that, because of the vulnerability of the client and the position of power of a counselor, it is so difficult for clients to act on their own values that it is a duty of the counselor not to act in ways that conflict with clients acting on their own values—for example, by imposing her values upon a vulnerable client. It is, therefore, a violation of a professional duty to direct a client to make a particular decision.

A primary *goal* in genetic counseling is that the client make a fully informed decision, one that is based upon her own values and full information about her options, and nothing else. A genetic counselor is required to provide an adequate amount of accurate information relevant to the genetic decision to be made and has a duty not to interfere with the client making such a decision. For example, there are thousands of articles and volumes written about Down syndrome. When a client asks a counselor about Down syndrome, she is expected to give a reasonable amount of up-to-date information. A counselor who said, "I don't know anything about it," would, at a minimum, be viewed as incompetent and would be derelict in her duty. However, no counselor would be expected to give every bit of information known about this condition. She is encouraged to give the maximum amount of information that will be useful for the client within the normal time constraints typical in genetic clinics. She is also required not to offer her own personal views about Down syndrome.

This fundamental ideal of genetic counseling, that a client make fully informed decisions based solely on her own values, generates several goals, all of them related to preventing those circumstances that may result in the client not making a decision based only on full facts about the situation and on her own values. There are many obstacles to fully informed decision making. The information that clients must comprehend is often abstruse and highly technical. The client is often emotionally distressed and has great difficulty in thinking with normal clarity. The client may be under pressure from a spouse or relative to take a certain course of action.

Valid decision making requires that the client be provided with adequate information, not be coerced by any member of the health care team, and be competent to understand and appreciate the information given. But it does not require that the client actually use that information in making her decision. It also does not require that the client be in the kind of mental and emotional state that will be most conducive to her making a decision based upon the facts and her own values. Finally, it does not require that there be no coercion by family or friends. It is acting on the ideals of the profession to try to assist the client to make a fully informed decision, to help the client to actually use the information provided to make her decision, to help the client achieve an emotional state that is conducive to good decision making, and to seek to mitigate or eliminate the coercion of family or friends.

A counselor has a duty not to impose her own values on her clients. It is acting on an ideal of the profession to help the client make a decision that, given all the facts, best represents the client's own values. A counselor has a duty to provide accurate information; it is acting on an ideal to try to help the client use that information in making her decision. A counselor has a duty to not coerce the client into making any particular decision; it is acting on an ideal to mitigate or eliminate the coercion of family and friends. A counselor has a duty to assess the competence of the client so that a competent client is permitted to make her own decision and an incompetent client is provided with a guardian to make decisions in her best interests; it is acting on an ideal to help the client achieve an emotional state that is conducive to good decision making when making her decision.

Ethical Dilemmas and Nondirectiveness

The overall ideal of the counselor is to prevent a client from making an irrational decision, not merely one that would be irrational for anyone, but one that would be irrational for those having the values that the client has. Acting on this ideal often requires the counselor to make clear and forceful statements, sometimes even to challenge the decisions of the client. However, because such statements are directive in the sense that they direct the client to look at a certain set of facts or suggest to the client that she appears to be making an irrational decision, some counselors hold back from intervening because of a mistaken belief that this would involve violating their duty not to impose their own values upon the client. This holding back is done in the name of nondirectiveness. For example, we recently interviewed a group of genetic counselors. We asked them to describe ethical dilemmas that arise in their work. They identified a recurrent theme: dilemmas regarding nondirectiveness.

Here is a case from our field notes. A counselor stated:

> A client came to me for counseling. She was trying to decide whether or not to have an amniocentesis. I went over the information regarding her risk of bearing a child with Down syndrome and the risks of inducing an abortion through the amniocentesis. She told me, "I'm going to wait and see if I like the doctor. If he is warm, I'll have it done, if not I'll forget it." I was uncomfortable with the way that she was making this decision but I didn't want to be directive so I did not say anything.

The counselor believed that her duty required her to be nondirective. However, at the same time, she felt uncomfortable because a primary goal of genetic counseling is to assist clients to make fully informed decisions based on their own important values. Thus she believed that she was caught in a moral dilemma. She is not alone among her colleagues in experiencing this misunderstanding.

The fundamental value that genetic counselors place on nondirective counseling is often repeated in the current literature in this profession. For example, a recent issue of the newsletter of the National Society of Genetic Counselors (NSGC) published the views of three prominent genetic counselors. They were asked how they would respond to this question from a client: "What would you do if you were in my place?" Each of the respondents begins with a salute to the value of nondirectiveness.

- "Nondirectiveness has been the hallmark of genetic counseling since early in the formalization of the field" (Godmilow 1990).
- "The essence of genetic counseling lies in the principle of autonomy, suggesting that our patients have the basic right to make reproductive decisions free from coercion. When we define the process of genetic counseling, we wave the banner of nondirectiveness, and rightly so, as it is derived from the principle of autonomy" (Edwards 1990).
- "Among the many arguments against directive genetic counseling, one of the most significant is that it can potentially violate one of the basic tenets in medicine—that is, above all else to do no harm" (Polzin 1990).

There are several reasons why nondirectiveness is held in such esteem in this profession. First, the nature of the decisions that patients bring to genetic counselors are both momentous and extremely private. They concern reproductive rights and often include the decision of whether or not to allow a fetus to come to full term. Most observers would agree that such profound decisions should not be made lightly or in any way imposed upon a patient. Surely, a person who must make a reproductive decision would not want, even in retrospect, to feel that she had made a decision that was imposed from outside. It would seem that, if society is to respect an individual's freedom in any area of life, it would be that of reproductive choice when there is the possibility of a genetic malady.

Related to respect for an individual's freedom to choose in regard to these matters is the value placed upon pluralism in a multicultural society. The code of ethics for the National Society of Genetic Counselors (NSGC, 1991) states,

> The counselor-client relationship is based on values of care and respect for the client's autonomy, individuality, welfare, and freedom. The primary concern of genetic counselors is the interests of their clients. Therefore, genetic counselors strive to: . . . Respect their clients' beliefs, cultural traditions, inclinations, circumstances, and feelings.

In a multicultural, secular society that shelters a wide array of religious beliefs and traditions, tolerance is of primary importance for preserving civility and protecting the rights of minorities. A genetic counselor who demanded that Catholic clients obtain abortions for fetal anomalies would correctly be regarded as intolerant and therefore would be likely to be repudiated by the profession.

Informed Decision Making and Nondirectiveness

Another reason that the profession places such high value on nondirectiveness is the central place of informed consent as a product of genetic counseling. The NSGC (1991) Code of Ethics states, "Therefore, genetic counselors strive to: . . . Enable their clients to make informed independent decisions, free of coercion, by providing or illuminating the necessary facts and clarifying the alternatives and anticipated consequences." The profession sees itself as a facilitator and steward of informed decision making. Directive counseling presumably alters this role in a way that would conflict with the goal of having the client make a decision based solely on her own values.

In the particular circumstances that surround genetic counseling, genetic counselors have good reasons to be concerned about being able to achieve their goal of having the client make a decision based solely on the relevant information and her own values. Two aspects of the context are significant. First, many of the decisions that patients must make take place in an atmosphere of crisis. For example, decisions to undergo an elective abortion often must be made under extreme time pressure. There are legal or practical limitations on abortions past the first trimester. Some key genetic tests can be performed only late in the first trimester and often require from 1 to 2 weeks before results are available. Consequently, the time interval between receiving a prognosis and deciding about abortion is short. In addition, the issue is a highly emotional one that often has tragic overtones for patients. Under these circumstances, there is concern that a counselor may have undue influence because the patient is in a vulnerable emotional condition. The patient must make momentous decisions at a time when calm reflection may be difficult if not impossible.

Further, a patient may need to be protected against undue influence from the medical environment, which forms the context for most reproductive decisions that pertain to genetic counseling. The men and women who must make decisions that fall under the purview of genetic counseling face a powerful social environment. Counseling usually takes place in a medical context (i.e., a hospital or clinic). More often than not, they are referred for genetic counseling by a physician. In a medical context there are prevalent values that pervade decision making in general. Biophysical phenomena that deviate from the norm are usually labeled as pathological and are usually made the target of ameliorative action. People who seek genetic counseling are often referred to as patients. Counselors are often nurses or physicians. The information that consumers of genetic counseling seek is often highly technical; the language is certainly outside the normal realm of daily discourse for most patients and thus often difficult to understand. In this circumstance, the counselor is in possession of a large body of technical information that makes her an expert. The patient, in contrast, is likely to have only a rudimentary knowledge of the biological and statistical issues that arise in genetic counseling. Consequently, the patient is in a vulnerable state in which he or she is relatively disadvantaged compared with the professionals in the medical setting. Thus, Clarke (1991) claims that a decision to refer a person for genetic testing is tantamount to a decision to abort in the event of a positive test result. He argues that, "Ostensibly, nondirective counseling in connection with prenatal diagnosis is inevitably a sham, not because of personal failure on the part of the genetic counselor but as a direct result of the structure of the encounter between counselor and client." In the light of the potential for undue influence by medical professionals, nondirectiveness seems to provide the extra safeguards against violations of the client's freedom that are required.

Problems with Nondirectiveness

Our quarrel with the "banner of nondirectiveness" is not that it is never justified. It seems clear that there are ample reasons for trying to avoid influencing what decisions vulnerable clients make. These are extremely personal and momentous decisions in a setting that entails powerful symbols and implied preferences. Rather, the problem with the emphasis given to nondirectiveness is that a mistaken or confused adherence to it can lead a counselor to avoid helping clients when they are making decisions that are clearly irrational, given their own values. As our case example illustrates, some counselors adhere to nondirectiveness *even in the face of decision-making processes that are recognized by everyone as flawed.* When taken to its extreme and regarded as a rigid and invariant duty, nondirectiveness can, in fact, work against the client making a decision that best reflects her own values.

In order to be useful, nondirectiveness or the principle behind it should enable one to discriminate between those cases in which nondirectiveness is appropriate and those in which it is not. When examined closely, nondirectiveness, like any other simple slogan, proves inadequate as a guide capable of unambiguously prescribing action in ethically complex situations. To begin with, it is virtually impossible for a professional whose job involves acquiring and sharing of technical information to remain radically nondirective and, at the same time, to fulfill the goal of counseling—facilitating a patient's informed decision making. A counselor has a duty to provide all the information that clients need in order to make rational decisions based upon their own rankings of the harms and benefits involved. The counselor must decide what information their clients need, how to present the information, and how to determine whether or not a patient understands the information. By virtue of her professional training and role, the counselor is expected to make these decisions without involving the client. For example, most would agree that it would be inappropriate for a genetic counselor to say to a patient, "I can tell you the good news about this genetic condition, or I can tell you the bad news, or I can tell you both. Which would you like to hear?" A competent counselor would make the decision about what information is relevant and would attempt to provide all of the information that a rational person would want to make the decision. But notice that, when the counselor assumes the responsibility for selecting the information to provide, she is acting in what some might claim is a directive fashion by editing the information and by urging the client to accept it.

Another example further illustrates the point that there are occasions in which a radical adherence to nondirectiveness conflicts with the counselor acting on the ideal of helping the client to make a fully informed choice. In this case, the patient says to the genetic counselor, "I have already made up my mind. I am not going to give birth to a monster. If my child has Down syndrome I will abort." In this case, the client has a false belief about persons with Down syndrome. The counselor in this instance seems faced with two competing tenets, nondirectiveness and the duty to facilitate informed consent. In the language of the NSGC Code of Ethics, on one hand the counselor should respect the client's "inclinations and feelings" and on the other she must "strive to enable the client to make an informed decision by providing the necessary facts. . . ." In order to provide the counselee with sufficient information to permit her to make an informed choice, the counselor will have to provide information that runs counter to the client's inclinations and feelings. Clearly waving the banner of nondirectiveness either does not lend much guidance to the counselor in this instance or guides her in the wrong direction.

Nondirectiveness and Autonomy

If rigid adherence to nondirectiveness can prevent a counselor, in some cases, from acting on the ideal of facilitating informed decision making, how is a counselor to know when to be directive and when to be nondirective? Perhaps a better understanding of the rationale for nondirectiveness would provide guidance. Some of the authorities quoted previously base their trust in nondirectiveness on the principle of preserving the client's autonomy. Thus, perhaps an examination of the concept of autonomy would provide guidance as to when a counselor should be directive and when not. In the discussion of principlism in Chapter 3, we examined the concept of autonomy and showed its inadequacies. Thus it should not be surprising that, because the tenet of nondirectiveness rests on the principle of autonomy, it has also proved to be inadequate.

Beauchamp and Childress (1983) provide an extensive discussion of the principle of autonomy. They first define the concept of autonomy as ". . . a person is autonomous if and only if he or she is self-governing. Autonomy as governance in the absence of controlling constraints points to the individual able to legislate norms of conduct (Kant) and able to voluntarily fix a course of action (Mill). Only if these conditions are present is a person autonomous." But as we have seen before, Kant has a very different concept of autonomy from that currently used in most medical contexts. Beauchamp and Childress go on to state a principle of autonomy based on this concept: *Autonomous actions and choices should not be constrained by others.* For all practical purposes, this principle says no more than the moral rule that prohibits depriving a person of freedom. Both the principle and the rule are regarded as prima facie guides; that is, they must be followed unless one has adequate reasons that permit one not to do so. Beauchamp and Childress go on to describe the limitations of the principle of autonomy:

> The principle of the respect for autonomy thus neither determines what a person ought to be free to know or do *on balance,* nor what is to count as a valid justification for constraining autonomy.

By not embedding the principle of autonomy in the moral system, they provide no procedure for determining whether a given reason is adequate to justify constraining autonomy, or depriving of freedom. Furthermore, the principle of autonomy has the added disadvantage that there is considerable controversy concerning what counts as an autonomous action or choice.

Our purpose in examining the principle of autonomy that supposedly lies behind nondirectiveness was to see if we could garner some guidance as to when the tenet of nondirectiveness can be overridden. However, an examination of the principle of autonomy simply leads us back to the same dilemma—that a person's autonomy ought to be respected but that there are some conditions in which it can be constrained. It does not, of itself, illuminate these conditions. Waving the banner of nondirectiveness does not get us very far in making practical decisions and changing the focus to the underlying principle of respect for a client's autonomy does not get us any further.

Indeed, it seems clear that, as vague as the concept of autonomy is, answering a client's question about what you would do if you were in her shoes is not a violation of autonomy. It is certainly not a constraint on autonomy in any normal sense of constraint. It may not be promoting a client's autonomy to answer that question when you know that this will result in her not making an independent decision, but rather in her simply accepting what you said, but this is not the same as not respecting a client's autonomy. How then does respect for autonomy tell a counselor what to do when a client asks, "What would you do in my shoes?" Further, insofar as respect for autonomy tells a counselor to say nothing when a client states that she is making a momentous decision based upon factors such as whether or not the doctor is warm, it seems to offer the wrong guidance. If the tenet of nondirectiveness and its underlying principle of respect for autonomy are not useful guides, perhaps there are other guides that are of more practical use.

Valid Consent versus Informed Consent

We use the phrase "valid consent" as the term referring to the basic legal requirement, and we use the phrase "informed consent" as the goal of genetic counseling. In this usage, valid consent and informed consent differ in the following ways. Valid consent requires only that the client be presented with adequate information and be able to understand and appreciate it; informed consent involves the patient making use of this information and basing her decision on it. Valid consent requires only no coercion by the health care team; informed consent involves no coercion by anyone, including family members, so that the decision is genuinely based on the client's values and the information provided. Valid consent requires that the client be competent. Informed consent involves the patient being sufficiently reflective so as to make decisions that are consistent with her other major values. Thus a counselor should strive, as an ideal, to point out to a patient when her decision making appears to be highly influenced by extreme emotion or is characterized by irrational thinking. As an example of this latter point, the counselor is under no legal or moral duty to tell an agitated patient to take a few days to make a decision; however, a counselor who follows an ideal of promoting decisions in a relatively "cool moment" with the goal of helping the patient arrive at a decision that is consistent with her values would make efforts to calm a patient or encourage a period of reflection. This gives genetic counselors an ideal that goes beyond providing the minimum basic legal requirement for valid consent and tells them, where possible, to make sure that the consent is informed consent—that is, based on the client's values and information provided and free of any coercion.

The concept of informed consent offers a better source of guidance to counselors than nondirectiveness. There are two useful ways to view informed consent: (1) as an outcome of effective counseling and (2) as a process that structures the relation between the counselor and the counselee so as to result in an independent rational decision based on the client's own ranking of values. To a large extent, the purpose of genetic counseling is to provide a counselee with adequate information to make a rational decision based on her own values. As a result of effective counseling, a counselee should be able to make an informed decision regarding his or her specific reproductive concern.

In order for consent or refusal to count as valid or informed, the following conditions must be satisfied. First, the person giving the consent must be given adequate information. Informed consent also requires that she use this information in making her decision. Second, she must not be coerced by any member of her health care team. Informed consent also requires that she not be coerced by anyone. Third, she must be competent to understand and appreciate the information provided. Informed consent also requires that she be as competent as possible in making her decision.

Regarding adequate information, a substantial portion of the genetic counselor's professional role is that of an educator. The counselor possesses a body of knowledge that is not normally available to lay persons without intensive study. This information is often highly technical and is stated in the professional literature in a vocabulary that is obscure to the average person. The information base for the profession grows at a rapid rate and it requires considerable time and effort to maintain an up-to-date knowledge of the field. One job of the genetic counselor is to function as an intermediary between the technical body of knowledge and the lay person. In effect, she functions as an educator. She makes technical information intelligible. Otherwise the patient would be unable to make either a valid or an informed decision, because she would not understand the necessary information. The genetic counselor serves as an explainer, translator, and interpreter of what is, in effect, a foreign language. The information is made intelligible in order to enable the patient to make a rational decision based upon all the relevant facts that are available.

A second requisite of valid consent is that it must be free of coercion by any member of the health care team. Just as a decision made under false pretenses is not valid consent, because it is not based on adequate information, so consent given because of explicit or implicit threats due to the health care system is not valid, because it is not free of coercion by those requesting the consent. Coercion involves the threat of force or the actual use of force. Informed consent implies a higher standard in regard to undue influence on a patient's decision making. Thus a counselor obtaining informed consent ought to be prohibited from engaging in improper social interactions such as badgering or humiliating a counselee in order to gain her consent. Counselors should also strive to avoid overpowering or intimidating a patient through the use of arcane jargon, displays of authority, and so forth.

Because all interpersonal interactions involve some mutual influence, it is never possible to simply remove the impact of the counselor. Thus there will always be some ill-defined "gray area" in which rational persons would disagree over whether a given mode of communication involves undue pressure or manipulation. It is in this "gray area" that counselors appear to fall back upon the "banner of nondirectiveness" as a guide. There is a legitimate fear that the context and structure of genetic counseling can place the patient at such a disadvantage that, given the vulnerable condition of the patient and the unequal access to knowledge and power between the counselor and patient, advice-giving may function as the giving of commands. Thus consent would not really be informed but would rather be simply a response to undue pressure. It is a failure to meet an ideal when a counselor is manipulative or exceeds normal bounds of interpersonal respect in ways that fall short of outright coercion but that must necessarily remain somewhat vague.

There are different kinds of advice-giving. An important distinction separates unasked-for advice from requested advice. It seems far more plausible to regard the former as undue influence than the later. Imagine a counselor who begins a first counseling session in the following way: "I am an expert about genetic conditions. I want to tell you as clearly and directly as I can that spina bifida is a terrible condition. You must abort this fetus. You would be very wrong to have this baby." Most observers would rank these statements as improper acts of undue influence.

But what about asked-for advice? Here is the territory where we first began. Providing asked-for advice is not an act of coercion, but it could still constitute undue influence. On the other hand, responding to a request for advice might be considered to be a proper part of the counselor's role as an expert and a teacher. The reason that providing asked-for advice appears not to be an example of undue influence is that, by asking, the patient appears to give consent for the counselor to be somewhat directive, thus taking away any sense that the counselor is intruding upon the patient. However, one could still believe that, even if asked for, advice-giving will result in undue influence and hence not facilitate informed independent decision making. It is important to notice here that the concept of informed consent provides useful guidance in this ambiguous area. It helps us in a way that the concepts of nondirectiveness and autonomy fail to do. In fact there is more utility to be mined from the concept of informed consent. It's utility is enhanced by considering it a process as well as a product.

Informed Consent as a Guide

Recognizing that it is following an ideal for a genetic counselor to facilitate genuine informed consent shows the importance of another distinction regarding giving advice, whether asked for or not. This is the distinction between giving advice in order to provide the patient with a decision and giving the advice in order to *model or facilitate the decision-making process*. In the former, the counselor tells the client what decision to make. In the later, the counselor tries to help the client make her own decisions; for example, she provides an example of the decision-making process in order to illustrate for the client what factors to weigh and what is involved in rational decision making. The following dialogues illustrates the difference in these two points.

Example 1 Giving advice.

PATIENT: What would you do in my shoes?

COUNSELOR: I would have the abortion as soon as possible.

Example 2 Modeling the decision-making process.

PATIENT: What would you do if you were me?

COUNSELOR: This is such a personal decision that I don't feel comfortable telling you outright what I would do because I'm a really different person from you. But I can tell you the things that I would think about. From what you have told me, I would give a lot of thought to how I would handle having a child with a cleft palate. I would think about how much time I have for helping him or her with the condition. I would also think about how easy or difficult it was for me to get pregnant. I would probably want to look at some pictures of children with cleft palate and learn about the kinds of surgery that are available nowadays. I probably would want to talk to a parent of a child with this condition or to a person with cleft palate to see what kind of challenge it poses for people. In my own case, I would think that I could give the time and energy and money needed to take care of my baby's condition. I have known some people with cleft palate and I know that it has caused them some pain in the form of teasing from other kids and social discomfort, but I also know that there are some really good surgeries available and that the chances are pretty good that my child would not be badly disfigured. I would also be influenced by whether I thought that I might not be able to get pregnant again and, depending on how important it was for me to have a child, I would weigh these factors differently. I am not advising you what decision to make, but I am suggesting to you how I would think about it.

In the second instance, the counselor provides an answer in order to model the decision-making process. This approach carries less risk of imposing undue influence on the patient. However, it can be improved upon by making further use of the concept of informed consent.

When considered a process, informed consent offers considerable guidance to the counselor. Here is how informed consent may be viewed as a process: At appropriate times in the interaction between a counselor and a patient, the counselor should make clear that *the counseling process itself* has possible harms and benefits and allow the patient to decide whether or not to proceed with the interaction. Whenever empirical information exists about the effect of a specific kind of information or specific form of counseling, the counselor should include it in the list of harms and benefits. The following dialogue illustrates the use of informed consent in the counseling interview.

> COUNSELOR: My job is to give you information about your genetic make-up and to give you as much information as I can to help you decide about whether or not you should have more tests done on the fetus that you are carrying. You should know that sometimes people get very upset when we give them this kind of choice. I have had mothers tell me "I wish I had never started this genetic counseling in the first place. It would have been better just to not know and to take our chances." On the other hand, the kinds of tests that we can do now can really tell you a lot about the chances that your baby will be healthy and it can make a big difference to make a decision based upon getting this information. Should I go ahead and start telling you about the tests and the conditions that we are concerned about?
>
> PATIENT: Yes, I understand that this decision is going to be hard. But I would much rather know what I am getting into than stay ignorant.

In this interchange the counselor sought consent to continue with the counseling process. The counselor attempts to make the process of counseling as transparent as possible by labeling what she is doing at each stage and by offering the patient the choice of continuing or not at each important juncture. When the counselor has information about possible benefits and harms of the counseling process itself, she states them before proceeding. It is important to note that information about benefits and harm of the counseling process may be derived from the counselors own professional experience. Thus an experienced counselor might say, "In my counseling practice over the past 5 years, I have noticed that some people have trouble understanding probability. I have several examples of probabilities in normal day-to-day experience that I can share with you if you like. Sometimes people don't want more information, they feel as if it puts them on overload or they feel as if I am talking down to them. Would you like to take a few minutes and talk about probability?"

The use of informed consent as a process is useful in the instance of asked-for advice. The counselor can attempt to confront the issue of undue influence. She can tell the patient that customarily genetic counselors do not give directive advice, because they want to be sure to support the patient's own ability to make independent decisions. The counselor can further point out that there is a risk in giving advice that the patient will feel pressured to go along with the counselor. The patient can then give a more informed consent to continuing with a more directive counseling process.

Informed consent as a process provides some guidance to the counselor. However, it does not provide an unambiguous guideline and should not be used as a slogan or "banner to wave." There are some dilemmas in counseling that do not have any simple answer or, in many cases, any unique right answer. For example, our first case concerned a patient who appeared to be making a decision about whether or not to undergo an amniocentesis based upon her feelings about the physician's demeanor. Here are several alternative responses that a counselor might make. They are ordered along a continuum ranging from completely nondirective and unobtrusive to highly directive and intrusive.

Option 1

PATIENT: I am going to wait and see if I like the doctor. If so, I'll have the amniocentesis. If he is not warm, I won't.

COUNSELOR: O.K. It's your decision to make in the best way that you see fit.

Option 2

PATIENT: (Same as in option 1.)

COUNSELOR: So, you are going to wait to meet the doctor and decide based on whether or not you like him.

PATIENT: Yes, that's right.

COUNSELOR: It sounds like the relation with the doctor is really important to you. It is so important that you are going to give it the most weight in making your decision.

PATIENT: Yes.

Option 3

PATIENT: (Same as option 1.)

COUNSELOR: It sounds as if the way the doctor treats you is really important.

PATIENT: Yes.

COUNSELOR: In my experience, women often find the idea of having a test that involves a big needle to be scary. If I were feeling scared about a medical procedure, it would help if the doctor was warm and friendly.

PATIENT: Yes, that's what matters to me now.

COUNSELOR: O.K.

Option 4

PATIENT: (Same as in option 1.).

COUNSELOR: It sounds as if it matters to you a lot that the doctor is kind to you.

PATIENT: Yes.

COUNSELOR: I sure understand your feelings. If this doctor is not someone you like, I can help you to find another.

PATIENT: That would be helpful.

COUNSELOR: It worries me a little bit to think that you might make your decision based upon how the doctor talks to you. We know that you are at high risk for having a child with a genetic problem. If you decide to skip the test because you do not like the doctor, then I worry about how you will react in the event that your baby has a disability. I believe that it is important for you to think about whether or not you want to have this child regardless of a disability. I believe that we can find a doctor who will be kind to you so that you don't need to worry about that side of it. If we take care of your concern about the doctor's manner, then what do you want to do?

These examples represent a progressively more directive counseling intervention. The final option, 4, represents a case of unasked-for advice-giving. It is clearly a violation of a radical nondirective approach, although it is not clear that it involves any constraints on autonomy. Indeed it seems equally plausible to regard it as an attempt to promote autonomy. Nondirectiveness, even in the context of genetic counseling, should not be taken as an absolute duty. There is no commonly held moral rule prohibiting giving advice, even unasked-for advice; nor is it necessarily in conflict with the counselors' goal to facilitate genuinely independent informed consent. There is a justified concern in genetic counseling regarding undue influence, but this is undue influence on the content of the decision. The genetic counselor should try to influence the client to think clearly about her decision and to make it based on the best available information together with a clear understanding of her own values.

Considering Public Policies

Rational people can clearly disagree about whether or not intervention 4 is done in the most appropriate manner. It may seem too direct or condescending. However, rational people should not disagree about whether or not the goal of intervention 4 is justified. It seems clear that it is an attempt to influence not the content of the decision but only the method of arriving at it. Any disagreement is *not* a debate about whether this intervention is a violation of a moral rule prohibiting depriving of freedom or a violation of the principle of autonomy. It is a practical debate about whether or not in seeking the goal of the genetic counselor to facilitate independent informed decision making, interventions as strong as those in option 4 should be used. That is, a counselor could chose option 4 and not be guilty even of a violation of the highest ideal of a genetic counselor. A test of whether this kind of unasked-for advice is acceptable is to ask whether or not it could be *publicly allowed*. In other words, might a genetic clinic post a sign that reads as follows:

> Our purpose in offering genetic counseling is to provide up-to-date information and to assist patients to make a genuinely independent and informed decision based on their own values. In the event that a counselor believes that a patient is not making this kind of decision, the counselor may choose to point this out and to give unasked-for advice.

Indeed, those who feel that this should be the policy of a genetic counseling center might regard it as part of a policy of informed consent that the center does post such a notice. Although it would now probably be considered odd to post such a notice, not only is there no moral rule against doing so, but it seems to be required by the doctrine of informed consent. Indeed, if one supported a radical nondirective policy, informed consent would require that a public notice be posted advising clients of that policy.

> Our purpose in offering genetic counseling is to provide up-to-date information and to assist patients to make a genuinely independent and informed decision based on their own values. However, in the event that a counselor believes that a patient is not making this kind of decision, the counselor will not point this out and will never give unasked-for advice; a counselor may even refuse to give asked-for advice.

Different well-intentioned people might disagree over which of these policies to adopt, but we think that most people coming in for counseling, if given the choice between the policies represented by these two notices, will choose the first. This does not answer all the questions about the relative weight to assign to respect for a client's predilections and feelings versus concern for the client's making a momentous decision, using the best possible methods. The directive option certainly seems as morally acceptable behavior as the radical nondirective approach.

Conclusions

The tenet of nondirectiveness proves to be a rather unreliable guide. The question whether to give unasked-for advice and under what conditions is not one that can be resolved by application of a simple slogan. Nondirectiveness and directiveness are both morally allowable approaches to counseling. Consequently, we recommend that genetic clinics should establish internal review processes that permit counselors to discuss cases with one another and to develop local norms regarding questions of directiveness. That is, the matter becomes a matter of professional judgment in which counselors should seek the support of peers through a process of case reviews. In this process, there are likely to always be some disagreements over the relative weight that different counselors assign to nondirectiveness versus more confrontational counseling styles.

Not only is there no simple slogan that can guide a counselor in how to respond in any given case, but in some cases there may simply be no best way to make a decision. The kinds of issues that genetic counselors face may often be essentially intractable dilemmas. In such cases, it might be just as good to toss a coin as to go through a tortuous decision-making process. It may well be that the patient who said, "I'll wait to see if I like the doctor to make my decision," is in effect saying "I will leave it to chance to decide." Given the tragic and unresolvable questions that often arise in genetic counseling, this kind of abandonment of rational decision-making processes may be a rational alternative. This argument flies in the face of the primary value given to informed decision making by the profession. It may be useful, however, to recognize that the goal of informed decision making is, in some cases, not really attainable because there are often so many unknowns and so many competing concerns.

Thus genetic counselors may need to recognize that there is considerable confusion in the first of the two primary tenets of the profession: nondirective counseling. We believe that this confusion is so great that the tenet ought to be given up. We think that the second tenet of facilitating informed decision making is far more useful than the first as a guide to the behavior of counselors. However, we are aware that even this second tenet has inherent limitations and does not provide an absolute guide. Sometimes, a patient who prefers to do so should be allowed to make her decision arbitrarily or by chance without disapproval from the genetic counselor. However, it is clear that, when informed decision making is appropriate, the counselor should use that degree of directiveness that will most facilitate independent and informed decision making by the client. We hope that this clarification of the duties and goals of genetic counselors, together with an explanation of the confusions involved in the tenet of nondirectiveness will be helpful to genetic counselors. We realize the great value of genetic counseling and we realize the need for a strong sense of humility in the face of extremely challenging moral dilemmas. We have simply tried to remove some of the confusions that have served as obstacles to counselors who strive to fulfill the ideals of the profession.

References

Beauchamp T.L. and Childress J. (1989) *Principles of Biomedical Ethics,* 3rd ed., New York: Oxford University Press.

Clarke, A. 1991. Editorial, *Lancet,* 338.

Godmilow, L., Edwards, J., and Polzin, S. (1990) Can non-directiveness be helpful: Three views. *National Society of Genetic Counselors Newsletter* 12(3):1.

National Society of Genetic Counselors. (1992) Code of Ethics of Genetic Counselors. *J. Genetic Counseling* 1(1):41–43.

7

The Concept of Genetic Malady

A person has a malady if and only if he has a condition, other than his rational beliefs and desires, such that he is suffering, or is at increased risk of suffering, a harm or an evil—namely, death, pain, disability, loss of freedom, or loss of pleasure—and there is no sustaining cause of that condition that is distinct from the person. In this chapter we examine what is meant by (1) harms or evils, (2) distinct sustaining cause, (3) rational beliefs and desires, and (4) an increased risk. We then develop the concept of a genetic malady and show how this concept may affect what counts as a malady.

Introduction

Knowledge acquired as a result of new genetic discoveries may change our views about persons and about the diseases or disorders from which they suffer. Instead of using the more limited terms "disease" and "disorder," we shall use the term *malady* as the most general term covering all of those states or conditions of persons that involve their suffering some harm or evil or having an increased risk of suffering a harm, for which there is no distinct sustaining cause. (We discuss the concept of a malady in more detail in the next section.) We believe that new genetic discoveries will increase both the number of persons known to have conditions already regarded as maladies and the number of conditions many will come to regard as maladies. Both of these increases may affect our attitudes toward ourselves and others whose genetic structures become more fully known.

It is likely that in the near future it will be possible to map a person's entire genome and sequence her individual genes. When this happens it is also likely that a great deal will be learned about the correlation between individual genes or combinations of genes and the occurrence of particular physical and psychological conditions, some of them maladies, some of them not. In some cases, the discovered correlation will be between a single gene (either a dominant gene as in Huntington disease or a recessive gene as in Tay Sachs), which may be only slightly modulated by other genes, and its physical or behavioral manifestations. But, in most other cases, the manifestations will be under polygenic control. In some cases, this correlation may be exact, with the environment controlling only the timing and the severity of the malady. Thus, the discovery of particular genetic sequences early in life may predict unerringly the appearance of a particular malady at some time. In other cases, the correlation between the genes and the malady will be lower, because some specific environmental circumstances will be necessary for its appearance, but knowledge of the genome will still allow better predictions than are now possible.

It is of more than theoretical interest whether a condition is regarded as a malady, and whether a person is regarded as suffering from a malady. Whether a condition is classified as a malady, for example, may determine whether it will be covered by Medicare or whether third-party payers will reimburse for its treatment. The malady status of some human conditions is noncontroversial: everyone would agree both that cancer is a malady and that having blonde hair is not. But what about infertility and alcoholism: are they maladies and should their treatment be covered by a national health insurance plan? More closely related to the problems that will result from the new genetic discoveries is the question: Is short stature ever a malady? If we can correctly predict that, without medical intervention, a boy will have an adult height of less than 5 feet, does he have a malady that should be treated and should treatment for his condition be reimbursed? And should his malady status be affected by the cause of his condition—for example, whether he does or does not have one or more hormonal levels out of the normal range?

Whether a condition is classified as a malady may determine more than whether there is reimbursement for its treatment. Some commentators, discussing the ethical justification of possible future genetic engineering interventions, especially germ-line gene therapy, have suggested that those that are directed toward the elimination of a malady are more justifiable than those that are not (see Chapter 10). Some argue that to use germ-line gene therapy to prevent or eliminate the malady of mental retardation would be justifiable, whereas to use it to ensure that one's offspring are more intelligent than average would not be. All of these issues raise the question of how the concept of "malady" is best defined.

The Definition of "Malady"

In order to discuss the ramifications of the new genetic discoveries on the concept of genetic malady, it is helpful to have a more detailed discussion of what is meant by a malady. Clouser, Culver, and Gert introduced the term "malady" as a technical term in order to have a broad generic term that included within it all of the narrower species terms such as "disease," "disorder," "illness," "injury," "lesion," "trauma," "headache," and so forth.[1] Using this term enables one to discuss all of the relevant issues without getting into semantic difficulties caused by the connotations of all of the more specific terms such as disease and injury. The definition of malady is as follows:

> A person has a malady if and only if he has a condition, other than his rational beliefs and desires, such that he is suffering, or is at increased risk of suffering, a harm or an evil (death, pain, disability, loss of freedom or opportunity, loss of pleasure) in the absence of a distinct sustaining cause.
>
> *Clouser, Culver, and Gert, 1981*

Although this definition may seem to be quite different from what might be expected, it is neither peculiar nor idiosyncratic. It accords fairly closely with the definitions of "disease" and "illness" that have been formulated by other authors.[2] Even more important it not only includes as maladies all of the clear cases (for example, cancer, broken limbs, and manic depressive disorder), but also excludes all of the clear cases (for example, hair and eye color). It also seems to resolve some problematic cases in an appropriate fashion—for example, it classifies alcoholism and infertility as maladies because they involve disabilities, but excludes normal grief because it is based on a rational belief. This definition not only captures the normative character of maladies, but also explains why people do not normally want to suffer a malady. It avoids the cultural relativity of other proposed normative definitions and is even more precise and objective than most nonnormative definitions. However, it still retains some vagueness, because, although we would not want to include as maladies those conditions that involved the suffering of very trivial harms or an insignificant increase in the risk of suffering a nontrivial harm, there may not be complete agreement on when a harm is sufficiently trivial or an increase in the risk sufficiently insignificant to lead us to refuse to count the associated condition as a malady. Assuming that there are always some costs or risks involved in preventing a condition from occurring or in treating it when it does occur, it seems plausible to count as maladies only those conditions that not only meet the definition, but for which, in the absence of any reason for doing so, it would be irrational to reject a low-cost and low-risk intervention. This preserves sufficient vagueness to allow for borderline cases and cultural differences while correctly and objectively classifying the overwhelming majority of conditions.

Four elements of this definition require explanation: (1) *harms* or *evils,* (2) *distinct sustaining cause,* (3) *rational beliefs and desires,* and (4) *an increased risk.*

1. *Harms* or *evils* (namely, death, pain, disability, loss of freedom, and loss of pleasure) have in common the fact that all persons acting rationally want to avoid them unless they have an adequate reason for not doing so. Harms or evil are the genus of which death, pain, and so forth, are the basic species. What death, pain, and the other harms or evils have in common is normative; that is, rational people want to avoid them. Maladies include the suffering of harms or evils, which explains why "malady" is a normative term. All rational persons want not to have any malady, unless they have an adequate reason for having some malady. On rare occasions, there are adequate reasons for having a malady; having a malady may enable someone to avoid having to do something (for example, join the army), which a rational person may rank as involving suffering more serious risks of harm than the risks or harms involved in having some particular disqualifying malady. In almost all cases, however, maladies are unwanted, and it would be irrational to want to have one.

2. Not everything that causes or increases the risk of suffering an evil is a malady. A person is suffering a malady if and only if the harm (or increased risk thereof) she is suffering is not in continuing dependence upon causes distinct from herself. A *distinct sustaining cause* is a cause whose effects come and go simultaneously (or nearly so) with its respective presence or absence. Thus a wrestler's hammer lock on a person may be painful but the pain is not a malady because, when the hammer lock ceases, the pain ceases. If the pain does persist later, then the hammer lock has caused a malady. Thus a malady is a condition of the person such that, whatever its original cause, it is now part of the person and cannot be removed simply be changing his physical or social environment.

3. We exclude *rational beliefs and desires* from the harm-causing conditions that count as maladies because no one holds that someone has a malady if he has a rational belief that causes him pain (for example, that his wife is dying of cancer) or if he has a rational desire that increases his risk of death and disability (for example, to go mountain climbing). A belief is irrational only if its falsity is obvious to almost everyone with similar knowledge and intelligence. We count as rational beliefs any beliefs that are not irrational. Irrational beliefs, called delusions by psychiatrists, are not excluded from the harm-causing conditions that count as maladies; indeed, irrational beliefs or delusions (for example, a belief that the government is trying to kill you) are often regarded as part of a malady, especially when they increase the risk of one's suffering some harm. As with rational beliefs, we count as rational desires all desires that are not irrational. An irrational desire is a desire for any evil or harm, or for what one knows will result in one's suffering a harm, when one does not have an adequate reason for that desire. Irrational desires

(for example, a desire to jump off tall buildings) are not excluded from the harm-causing conditions that count as maladies; indeed they are a major symptom of many mental maladies.

4. The new genetic discoveries make it very important to be clear about the concept of *an increased risk,* because one of the results of the new genetic information is that many people will find that they are at an increased risk of suffering some harm owing to their genetic endowments. When one is concerned with whether a condition associated with an increased risk counts as a malady, an increased risk means an increased risk over what is normal for the species, not an increased risk over what is normal for some group to which one belongs or what had formerly been normal for oneself. To have a condition—for example, the gene for Huntington disease—that makes one significantly more likely to suffer some harm than those without that condition, counts as having a malady only if the overwhelming majority of people do not have that condition, or at least did not have it in their prime. There is a complete parallel with the concept of a disability. Lack of an ability (or a low level of an ability) counts as a disability only if the overwhelming majority of people have that ability (have a higher level of that ability), or at least did have it in their prime. Given the enormous number of things that can go wrong with a person, it is very likely that all of us suffer from some genetic conditions (one estimate is at least five) that make it significantly more likely that we will suffer some harm than the overwhelming majority of the species who do not have those conditions.

"Malady" is meant to be a genus term, of which "disease," "disorder," "injury," and so forth, are species terms. Thus all diseases, all disorders, and all injuries are maladies in our account. It is conceptually useful to have one general term that captures what all the more specific terms have in common. One problem with searching for general definitional criteria in the more specific terms is that each term has certain connotations attached to it that are now regarded by many as arbitrary. For example, in ordinary English a broken leg is an injury and it is a mistake to classify it as a disease, although broken legs clearly satisfy most prior attempts to define "disease." Medical knowledge has shown that there is no important fundamental distinction between injuries and diseases and in fact some pathology textbooks refer unambiguously to injuries as "traumatic diseases." "Malady" is intended to provide a less misleading general term to refer to both injuries and diseases, as well as to mental disorders and headaches. Oddly, no language we have investigated, including English, has such a general term.[3] Therefore we brought back what we took to be an archaic term, but one that had a sense close to what we wanted, and gave it this new technical sense.

Our definition was elaborated to try to make precise what was common to disease, disorder, injury, etc. Our definition of malady is unique in that it acknowledges that malady is a normative concept, incorporating the basic normative concept of a harm or an evil as a necessary part of the definition, yet still recognizes that malady is an objective concept. We can do this because we recognize the objective character of the normative concept of a harm or an evil (see Chapter 2). There is considerable evidence that prior definitions of the concept of disease either tried to deny its normative features or based any included normative features on societal criteria. Thus they failed to capture an essential element of the concept or they were subject to influence by all kinds of subjective agendas, such as enforcing certain conceptions of morality or denying that some persons had maladies. By making the normative features explicit, we have been able to make them very precise, so that we can not only account for the normal understanding of the concept, but also make it both universal and objective, and so more resistant to manipulation for political or other purposes. Our definition makes clear that deviance or abnormality in and of itself, whether physical or mental, is not sufficient for a condition to be counted as a malady. Having an unusual laboratory value or an unusual sexual preference, unless these conditions are universally associated with the person suffering harms or evils or an increased likelihood of suffering harms, does not count at all as showing that the person has a malady.

The foregoing definition of malady has proved useful in the analysis of many conditions whose malady status was uncertain. For example, the definition makes clear that alcoholism qualifies as a malady because alcoholic patients do suffer from a harm that has no distinct sustaining cause. They have a volitional disability with regard to the drinking of alcohol: whether or not to drink alcohol or, once having begun to drink, whether to continue drinking is not a voluntary action for an alcoholic.[4] A slight variation of the definition of malady was used by the authors of the American Psychiatric Association's *Diagnostic and Statistical Manual* (DSM-III-R) to define mental disorders and to distinguish between them and physical disorders. That distinction was made on the basis of the salient symptoms of the disorder. Because the definition of malady has been useful, it seems justified on heuristic grounds to use it to define the concept of a genetic malady and distinguish genetic from nongenetic maladies. It is clear, however, that this distinction will not be based on the salient symptoms, but rather on the salient causes of the symptoms or harms suffered. A genetic malady will be a malady in which genetic rather than environmental causes are salient. However, it is important to remember that, for most maladies, both genetic and environmental factors will be involved.

What Conditions Are Genetic Maladies?

When, according to the foregoing definition, are conditions correlated with particular DNA sequences genetic maladies?

A person clearly has a genetic malady if he is directly suffering harms because of his genetic condition (e.g., he has Tay-Sachs disease) or because of his chromosomal structure (e.g., she has Down syndrome). Such conditions qualify according to the definition because harms are being suffered (e.g., mental and physical disabilities), and these harms are caused not by a distinct sustaining cause but by the person's genetic make up or his chromosomal structure

In contrast, there are many conditions that are genetically determined, fully or partially, but that are not maladies because they do not involve the suffering of harms. Eye color and fingerprint patterns are two clear examples. Having hazel eyes is not a malady. Neither is having each of all ten fingers with a fingerprint pattern of an arch (or having ten loop patterns, or ten whorls), though this would be statistically extremely rare. Although eye color and fingerprint pattern are genetically determined, nevertheless neither condition involves suffering or an increased risk of suffering harms. Therefore neither can be considered a malady, hence they cannot be genetic maladies, either intuitively or according to the foregoing definition.

It also follows from the definition that a person has a genetic malady if his genetic structure is regarded as being primarily responsible for an increased risk of suffering harms in the future. Huntington disease is a clear example: if a young woman discovers in her twenties that she has the *HD* gene, we would probably think of her as having a genetic malady even though she is not yet suffering any symptoms. The definition thus accords with our intuitions. The situation is similar to someone having a significantly elevated blood pressure but not yet having target organ symptoms or someone being HIV positive but not yet symptomatic with AIDS. The Huntington-positive and the HIV-positive person will certainly or nearly certainly suffer harms in the not too distant future. The person with the *HD* gene is even more likely than the hypertensive person to suffer harms in the future and nonsymptomatic hypertension is widely regarded as a malady.

It is possible that genetic testing will in the future reveal many conditions for which, if suitable prophylactic measures are taken, symptoms will not develop. Phenylketonuria (PKU) is a present example of this kind of genetic condition: if the condition is diagnosed sufficiently early and appropriate dietary precautions are followed, serious symptoms may be prevented. Many similar kinds of conditions may be discovered in the future in which a suitable diet, or the prolonged administration of a drug, may prevent symptoms from occurring. Even in the fortunate cases in which treatment results in the genetic malady's symptoms being completely avoided, affected persons would still be regarded as having a genetic malady. This is because they suffer the loss of freedom to eat certain foods, at least for some period of time. As long as one cannot eat certain foods or must take drugs chronically, then one continues to have the genetic malady. This situation is closely analogous to the condition of having an allergy. An allergy is a malady even if the person can eliminate symptoms totally by taking a drug or moving to another part of the country: her freedom has been curtailed by having always to take the drug or by having to live in one place and not another. Indeed, it is quite likely that allergies are genetic maladies. If the prophylactic measures in time are no longer needed, then the person would no longer have a malady; it is not clear whether to say that she has outgrown the malady or it has been cured.

Genetic maladies (e.g., Huntington disease and PKU) are far more likely than nongenetic maladies to include conditions in which harms are not currently being suffered but in which there is an increased probability, compared with the population at large, of harms being suffered in the future. Individuals with conditions such as Huntington disease may suffer no harms for a long period of time, but the harms, when they do appear, are quite severe. Sometimes, as in Huntington disease, the harms are certain to appear and there is no preventive treatment. Sometimes, as in PKU, the price of forestalling suffering those harms is to engage in treatment regimes that themselves involve the suffering of other nontrivial harms. It is for these reasons that it is in accord with our intuitions to consider persons with nonsymptomatic Huntington disease or those still being treated for PKU to be suffering from genetic maladies.

New genetic discoveries also may make available knowledge about genetic conditions that involve being at increased risk of suffering relatively mild future harms. Suppose we could learn, with a 50% degree of probability, that a four-year-old boy, because of some particular genetic condition, would suffer in his thirties from some mild skin condition, annoying but not disfiguring, such as eczema of the scalp. He would satisfy the definition of a genetic malady—that is, he would be at increased risk, compared with the general population—of suffering harms in the future because of a particular genetic condition, and it would be irrational to reject a low-cost and low-risk intervention. If there were no way to prevent those symptoms, it might seem questionable to regard him as having a malady at age four. Some may claim that, because the harms—even when they occur—are mild and because there is only a 50% probability of those harms occurring and even then not for thirty years, the condition is too trivial to be classified as a malady. However, if a low-cost and low-risk intervention were discovered that would in all cases forestall the eczema, it would be apparent that this genetic condition was indeed a genetic malady.

Whether we regard genetic conditions that involve possible or certain *future harms* as *genetic maladies* seems to be a joint function of several variables. Three seem particularly important: (1) the degree of probability that the harm will occur, (2) the seriousness of the harm if it does occur, and (3) the likely age of the person when it might occur. With regard to the first variable, the higher the probability of future occurrence, compared with the population as a whole, the more likely the person will be regarded, in advance of the harm occurring, as having a malady. Thus a 25-year-old with a 50% chance of suffering the harms of leukemia before age 60 might be regarded as having a malady but, if he were only from 1% to 2% more likely to develop it than others of his age, he probably would not. With regard to the second variable, suppose one has a genetic condition such that, if symptoms do occur, they would be sufficiently mild that it would not be clear whether or not to regard the person as having a malady (e.g., some cases of skin discoloration). Then a genetic condition that tripled one's risk of developing these symptoms at age 50—say, from 10% to 30%—probably would not be counted as a genetic malady. The age of occurrence (the third variable) may be the most important variable. If, on the basis of his genome, it could be predicted that in his nineties, a 25-year-old had a 50% chance of suffering the harms of leukemia, we would not regard him at age 25 as suffering from a malady. In fact, even if the likelihood of his developing leukemia approached 100% if he survived to his nineties, we still would not regard him at 25 as having a genetic malady. This is based on the assumption that the average life-span is in the mid-eighties or lower.

Someone having the *HD* gene would be regarded by everyone as having a genetic malady because of the seriousness of the symptoms, the certainty that they will develop, and the fact that death will occur prematurely. Huntington disease can be accurately diagnosed decades before any symptoms become manifest and the person may feel entirely well for decades. However, once diagnosed, no matter how early, the affected person would still be regarded as having a malady. Even though there is currently no treatment that may postpone or ameliorate its symptoms, if an expensive genetic treatment became available that would prevent the malady from developing symptoms at least 50% of the time, everyone would favor including it in health insurance coverage.

In the future, many significant linkages between genes and maladies may be found. Thus the number of persons who know or could discover at a relatively young age that they have a condition, without a distinct sustaining cause, associated with a significantly increased risk of suffering serious harms in the future could increase significantly. For example, the *age of occurrence* of heart disease and cancer, the two diseases with the highest death rates in the United States, will very likely prove to have significant genetic correlates that can be measured at an early age. This could result in a frequently noted ethical problem: the extent to which genetic information about individuals could and should be kept confidential and unavailable to employers and life insurance companies. But it could also result in a large number of young individuals, who are currently ignorant about their likely medical future, regarding themselves and being regarded by others as suffering from maladies because of the significantly increased probability that they will become seriously symptomatic at middle age.

The *age* of occurrence of future harms may be more important in determining whether a genetic condition counts as a genetic malady than either the *seriousness* of the harms or the *increased probability* of their occurrence. We are all mortal: the probability is 100% that all of us will eventually succumb to some malady. However if a person's genome contained no condition that resulted in an increased risk of developing a serious disease early in life, but he had a genetic condition that made it a near certainty, barring accidental death, that he would develop and die from heart disease in his mid-nineties, we would not regard him at age 25 as having a malady. However if, at age 25, his genome revealed that there was a 50% chance that he would die of heart disease before the age of 40, unless (or despite) engaging in vigorous treatment, we almost certainly would regard him as having a malady. Even if his expected age of death was in his sixties, he still would probably be regarded as having a malady.

Being at Higher Risk: Group Membership versus a Condition of One's Person Some individuals are known to be at increased risk of prematurely developing diseases such as heart disease and cancer because of their family histories. However knowing that a person is at increased risk because that person has a particular genetic structure associated with a high risk of prematurely suffering a serious malady and knowing that a person is at increased risk because of being a member of a group, some of whose members will prematurely suffer maladies whereas others will not, result in different perceptions of the malady status of two persons, both by themselves and by others. This may be true even when the objective level of risk is the same.

Consider this example: Jane is born into a family, 25% of whose female members develop breast cancer before the age of 40. We would ordinarily say of Jane, at age 20, that she is at increased risk, compared with other women, of developing a malady but not say of her that she in fact *has* a malady. However, suppose that Jill is born into some other family, none of whose female members has ever, so far as is known, developed breast cancer before the age of 40. We map Jill's genome and discover, with our new knowledge of correlations between genomic structure and cancer, that she has a 25% chance of developing breast cancer before the age of 40. We would be much more likely to say of Jill than of Jane that she has a malady. We would think, because Jill has an aberrant genetic sequence, that she has something wrong with *her*. Of Jane we would likely say that we do not yet know whether she has something wrong with her—that is, with one of her genetic sequences—even though her risk appears objectively the same. We would say that Jill has a malady but that we do not yet know whether Jane has a malady.

Our definition of genetic malady does distinguish between Jane and Jill. *Condition,* in the definition, means condition of the person. Jill is at increased risk because of an aberrant genetic sequence, and this genetic sequence is clearly a condition of her person. Jane, in contrast, is at the same increased risk, but we do not know if this is because of some condition of her person: being a member of a group (her family) is not what we mean by a condition of the person. If Jane does develop breast cancer, it is almost certainly because of an aberrant genetic sequence, as is true with Jill, but we do not yet know whether Jane has the aberrant genetic sequence.

If we are more likely to say, of two persons at equal objective risk of suffering genetically caused harms in the future, that the one with a demonstrable genetic aberrancy has a malady, compared with the other who only may have a genetic aberrancy, we can see the power that the new knowledge of linkages between one's genome and maladies could have in altering our perceptions of persons' malady status. Thus discovering genetic conditions of persons that significantly increase their risk of suffering future harms will result in many more people having maladies at a much younger age. By increasing the importance of having a significantly increased risk of suffering harms, the new knowledge will decrease the close connection between suffering symptoms and having a malady. Many more people will have maladies that have no symptoms. It may become much more important to distinguish between having a malady or a disease and being ill—that is, actually having symptoms or suffering harms.

More Conditions May Be Regarded as Maladies

As more and more is learned about the genetic underpinnings of various human traits, abilities, and physical characteristics, some conditions, which we now regard only as "normal" variations, may come to be viewed as maladies. Low intelligence and short stature are examples of this phenomenon.

Intelligence

All maladies involve the suffering, or the increased risk of suffering, harms. Among the harms that characterize maladies are various kinds of disabilities. Disabilities may be physical (say, being unable to walk), mental (say, being unable to remember one's name), or volitional (say, being unable to enter elevators). Most abilities are not simply present or absent; rather, different persons possess them to different degrees. Some persons, for example, are more intelligent than others, and fine gradations of various physical and intellectual abilities can be validly and reliably measured. However, not all individuals with abilities below the statistical average count as disabled, and thus as having a malady; on the contrary, in order to count as disabled, one must be sufficiently below the average that the overwhelming majority of people in their prime have significantly greater abilities. Further, in order to count as a malady, this low level of ability must not be capable of being remedied solely by appropriate teaching but must involve some medical intervention. In order to count as a malady, it does not matter whether the low level of ability was due solely to the condition of the person (e.g., one's genetic endowment) or to an early deprived environment; all that matters is that it is not now dependent on factors outside of the person (that is, his environment).

We now know that, except for those situations in which there is a serious genetic malady, intelligence is determined by a complex interaction between children's environment and their genetic endowment. The topic of the role of genetic factors in determining intelligence is highly controversial because historically claims about the genetic basis of intelligence have been used to justify eugenics policies that were clearly immoral. Persons with mental retardation have been subjected to unwarranted restrictions of their liberty. In Europe, persons with mental retardation and physical disabilities were among the first to be executed by the Nazis in Germany. The historical record concerning practices pertaining to genetics and intelligence is one that suggests widespread abuse and misunderstanding. There is a tradition of immoral state-sponsored practices associated with overly simple theories about the genetic basis of intelligence.

Many of the early successes of human genome research have centered on conditions for which there is a simple and direct causal relation between an abnormality in the human genome and a specific condition such as cystic fibrosis. No one expects intelligence to be determined by a single gene, but many still mistakenly believe that genetic factors alone determine human intelligence. However, complicated phenomena such as human intelligence rarely are determined by a single factor. Even if multiple genetic correlates of intelligence are identified, their relation to the child's environment will be a complex interactive one that is likely to defy the development of any simple prescriptive intervention. However, the history of eugenics suggests that people are eager and willing to believe that there are simple genetic causes for mental retardation and other behavioral problems—particularly when these characteristics are attributed to a disadvantaged racial or ethnic group.

For all of these reasons, a discussion of mental retardation as a genetic malady should proceed with great caution. We believe that a proper understanding of the concept of a genetic malady may help avoid some of the more serious misunderstandings. It must be made clear that mental retardation is a genetic malady only when (1) the retardation is sufficiently severe that the overwhelming majority of people in their prime have significantly greater abilities, which we take to occur when the IQ is below 70; (2) the person's genetic endowment is sufficient to account for this low level of ability, even if the level of retardation is worse because of a lack of appropriate teaching and could be raised to some degree by appropriate teaching. When the low level of intelligence is due primarily to environmental factors but cannot now be remedied by appropriate teaching, mental retardation is still a malady, but it is obviously not a genetic malady. As the following discussion makes clear, there may be a strong temptation to regard mild mental retardation (an IQ of 60 to 69), which has a strong environmental component, as a genetic malady.

The choice of 70 as a dividing line between malady and nonmalady is somewhat arbitrary, as is almost always the case when a line is drawn on a continuum, but the underlying point is clear: when IQ is extremely low, it constitutes a malady but, when it is merely below the average, it does not. It is only when a person's ability is extremely low (for example, an IQ below 70—more than two standard deviations below the mean) that we regard him as having a mental disability (for example, mental retardation). Although a person with an IQ of 80, for example, is less intelligent than well over half of the population, we do not regard him as being disabled or as having a malady. But someone with an IQ of 60 does have the malady of mental retardation. (DSM-III-R codifies this distinction: IQs below 70 are "Axis II Developmental Disorders," coded 317.00 or 318.00, depending upon their severity, whereas an IQ of 70 or above is not a codable disorder.) However, an IQ of below 70 counts as a genetic malady only when the genetic endowment of the person is sufficient to account for the IQ being below 70.

Intelligence is one aspect of a person's mental abilities. High intelligence can be used to further immoral or even irrational ends, but high intelligence in and of itself means that one has greater ability to take advantage of opportunities, to procure desirable ends, and to avoid harms. In ordinary circumstances, it would always be irrational to desire to be less intelligent. In all normal circumstances, it would be immoral to intentionally act in some way in order to have one's fetus grow into a child with a lower intelligence than would otherwise be the case. Nonetheless, although those with IQs of 80 are somewhat disadvantaged, it is a mistake to regard them as having maladies.

Increased knowledge of correlations between one's genome and various traits could alter some people's view of those with low but unretarded levels of intelligence. Suppose it is discovered that there are ten genetic loci that control most of the variance associated with intelligence. Suppose further that we can appraise these loci in utero and diagnose within 10 IQ points the eventual level of intelligence of a fetus (assuming no later cortical insult that lowers its intellectual potential). Imagine a man and woman of above average intelligence who determine, through fetal genome analysis, that the woman is carrying a fetus that will, when grown, be an adult with an IQ of about 80. This couple already has two bright children, they have little doubt that she will be able to become pregnant again, and the couple have no moral qualms about abortion. If they learned their fetus was clearly retarded (e.g., it suffered from Down syndrome), they would choose to abort it. However, it also seems possible that the couple would desire to abort a fetus with an IQ of 80. It would be understandable for them to want to do so; however, they may decide not to have an abortion if they regard an IQ of 80 as part of the normal variation, rather than as a genetic malady. Whether or not an IQ of 80 is regarded as a genetic malady also may determine whether or not a genetic counseling facility will offer a genetic test for determining that condition (see Chapter 9).

Even more speculatively, suppose that we could, with little risk, perform germ-line gene therapy on the fertilized egg at the ten genetic loci in a way that raised the fetus's potential from its probable IQ of 80 up to, say, a probable IQ of more than 100. If we distinguish between positive and negative eugenics, accepting that the latter is morally acceptable but the former is not, then deciding whether or not having an IQ of 80 is a malady may determine whether we allow such germ-line gene therapy (see Chapter 10). If we accept positive eugenics, it might happen that the majority of parents in this same situation would decide to have gene therapy to raise their fetus's intelligence in this way. This might mistakenly lead many to regard unaltered fetuses with IQs of 80 as having something wrong with them and to later regard the lower-intelligence adults into which they developed as having a malady. This would be a mistake, because even in the far foreseeable future not enough fertilized eggs could be given therapy such that it would affect whether, with regard to someone with an IQ of 80, an overwhelming majority of people have a higher level of ability, or at least did have a higher level in their prime. However, one effect of having a "treatment" for raising IQ might be to raise the threshold below which low intelligence was considered by many to be a malady. For example, although fetuses with a potential IQ of 90 would still be regarded as having a low-normal variation, those with potential IQs ranging from 75 to 80 might be regarded as having a genetic malady. Thus what is actually a normal variation might come to be regarded by many as a disability, and hence a malady, with an inevitable increase in discrimination.

Short Stature

Similar considerations apply to short stature. Even in the absence of endocrinal abnormalities, very short height would be regarded by many as a malady. For example a man in the United States who is four feet, six inches tall is likely to experience many harms because of his height: he is deprived of the opportunity of carrying out many physical activities available to men of average height and is likely to be the object of life-long social stigmatization of various kinds.

If a couple were informed, by genome analysis, that their fetus would grow into a man with an adult height of four feet, six inches, they might well feel that the fetus had a serious potential malady. The situation with regard to abortion is complex because, in contrast with that of potentially retarded fetuses, we have no tradition of aborting fetuses with potentially normal intelligence and normal life expectancy but with extremely unusual physical characteristics. However, if genetic therapy were available that would significantly increase the fetus's potential height while putting the fetus at little or no risk, it seems likely that the majority of couples who did not decide to abort the fetus would want the therapy to be given.

But consider a male fetus with a potential height of five feet, four inches. This is below the average height (about five feet, ten inches) of males in the United States, and probably the majority of males with this height suffer to some extent life long because of it, but it is clear that a man with this height does not suffer from a malady. However, as with the example of intelligence, suppose the majority of parents requested genetic therapy for fetuses with this potential height. It seems likely that over time men of this height would more and more be considered as suffering from a malady. In part this might be because men of this height would be seen less often and would thus represent greater and greater deviations from men's average height. However, in part it might be because once a condition that is not a malady, but is regarded by many as undesirable, can be corrected it is more likely to be regarded as a malady.

Thus, in the future, because of knowledge gained through the new genetic discoveries, more persons will be known to have maladies and more conditions may be regarded as being maladies than is now the case. The first change would not increase the number of maladies that occur in the world, but it would mean that more persons could know for a longer period of time that they suffered from a malady. The second change would increase the number of conditions that are regarded as maladies, and therefore the number of persons who think of themselves or are thought of by others, or both, as suffering from a malady. Even though we think that this second change is usually based on a mistaken concept of malady, it is likely to be widespread enough to affect the way people think about these conditions.

However, it may become possible to predict kinds of increased future risks that it would be counterintuitive and also pointlessly harmful to the individual to label in advance as a malady. As we suggested earlier, one kind of increased future risk does seem correctly labeled as a current malady even if the person is currently suffering no harms—namely, when there is a significantly increased risk that a serious malady will be suffered at a relatively young age. However other kinds of scenarios may be possible: the increased probability of a serious malady may be very slight or there may be a greatly increased risk of suffering very mild symptoms in the future. In addition, because one's genome can be appraised at any age, the time of probable appearance of the malady may not occur until after a normal life-span. Each of these variables is on a continuum and each can covary independently of the others.

Consequences of the New Genetic Knowledge

Maladies are, in and of themselves, bad things. No rational person wants to suffer harms or be at risk of suffering harms unless she has some reason, such as the belief that she will avoid suffering even worse harms. To discover that one has a malady, if it is serious, usually causes significant unpleasant feelings. When others learn that a person has a serious malady, they may view and act toward that person differently in a way that is harmful to him. Therefore any development, such as the new genetic discoveries, that increases the number of persons who think of themselves or are thought by others, or both, as having maladies is worrisome.

One way to minimize the resultant harm would be to stipulate, legally and through codes of ethics, that information about a person's genome is confidential. Like HIV testing, genome analyses could be done only with the valid consent of a competent person who would be furnished with counseling before and after testing. It would be discretionary whether an individual were ever tested. The information obtained by testing would be private, and there would be serious legal and ethical repercussions if confidentiality were breached by a practitioner. Health insurance companies could neither require genetic testing nor use information obtained through genetic testing. Thus each person could determine whether the possible harms that genetic testing might cause would be outweighed, for him, by the possible benefits gained, and each individual, if he were tested, could maintain full control over the information obtained. For this to happen, physicians must be clear about what counts as a medically private situation and on their duty to specify to patients in advance the boundaries of these situations before the patient allows private information to be generated or disclosed. The possibility of discrimination on the basis of genetic testing may be one of the reasons for the now overwhelming support for universal health insurance, which does not allow discrimination on the basis of preexisting conditions.

With regard to life insurance, the problem is more complex. It would seem that one cannot prevent life insurance companies from charging higher premiums to those who have not been tested or simply to refuse to insure those who have not been tested. In order to prevent discrimination by life insurance companies, one would have to prohibit them from even using the results of tests that the person tested wants them to use, for if such tests could be submitted in order to get lower premiums, then discrimination is possible even without anyone requiring genetic testing. Thus allowing individuals to use their own tests as they want can have serious effects on others. In many cases, such as health insurance, it is possible to avoid the conflict between an individual's right to use the results of her own genetic tests as she determines for herself and the harm caused to others who do not want to be tested. However, in many other cases, such as life insurance, allowing an individual the right to use the results of her own genetic tests as she determines for herself will result in harm caused to others who do not want to be tested.

There will always be circumstances when one would inform parents of prenatal genetic tests of fetuses—for example, to allow for abortion—if the results seem to warrant that. However, there seems less reason to examine the genome of children except to discover the possibility of their developing a serious malady that might be prevented or ameliorated through prophylactic treatment. The best general policy may be to defer the possibility of genetic examination until adulthood, when each person can make his own decision about the possible harms and benefits of learning about possible future maladies he might suffer. This does not settle all of the important social issues concerning the consequences of others knowing about one's genetic conditions, but it seems to give individuals greater control over what they will learn about their own genome.

That raises an interesting question: to what extent would individuals with no known risk factors want to learn about the probability or certainty of developing a serious malady later in life? The usual presumption is that many persons, perhaps most, would want to know the truth about any maladies from which they will suffer because, although the truth may be painful, it allows one to make realistic and appropriate plans for what lies ahead. The generally accepted view, confirmed by studies of those at risk for Huntington disease, is that although not knowing may be temporarily less unpleasant, in the long run one is better off with earlier knowledge. But this may not be true of genetic testing for those with no known risk factors.

Most of us probably plan our future life projects knowing that we are mortal but assuming, vaguely, that we will live until at least our mid-eighties (that is, for women, an average life-span), though perhaps hoping we will live somewhat longer. There is ongoing pleasure and utility in structuring our lives in terms of long-range and short-range goals and projects in accord with that average life expectancy. We know that we may die (or become seriously incapacitated) sooner: if mid-eighties is the average life expectancy, that means about half of all persons will die before then. We also know that the distribution of age-at-death is a fairly normal curve; so, if someone is in a cohort where the average age of death is eighty-four, we do not think she has died prematurely if she dies at seventy-nine or eighty-two, or perhaps even at seventy-five. But most women, in their twenties through their sixties, say, know no reason not to think of the mid-eighties, give or take a few years, as an approximate endpoint toward which to anchor their thoughts and plans.

If a woman learns at age twenty-five that she will probably or certainly die in her mid-forties, several related negative things could occur. Instead of planning for sixty more years of life projects, she can plan for only twenty. In addition, for quite a long while, perhaps even for twenty years, she might mourn her truncated life. She might feel envious and even resentful of others her age who, as far as they or she knew, had an average life expectancy. She might gain something: the ability to plan her affairs so that they drew to a sensible end after twenty years. For example, she might avoid starting long-term projects in her mid- to late-thirties, and so forth.

Of course, if she learned from her genetic assay that there was no indication that she would develop a major disease before her mid-eighties, that would be reassuring, but she had no particular reason to believe she would not live until her mid-eighties anyway, and so the gain in her well-being would be small. It is thus not clear, whether the results were positive or negative, that the possible gain would be worth the possible loss. This relative balance of losses and gains from early testing seems to apply homologously to close loved ones with whom she shared the results, negative or positive, of her genetic testing. Comparing the possible harms and benefits of testing, at the very least it seems rational not to be tested.

If early testing brought to light a serious condition that could be prevented or ameliorated significantly through early prophylaxis, there would be a very strong reason to be tested. However this seems unlikely, at least at present. Many or most serious maladies with a significant genetic component cannot be prevented or attenuated with present knowledge. Even if it could be learned that a person had a genetically linked likelihood of developing heart disease or cancer at a relatively young age, the kinds of prophylactic measures that might be suggested (diet, avoidance of carcinogens, frequent mammograms) might have an equally beneficial effect whether or not one had a genetically linked vulnerability. Thus, if a person had the kind of strong interest in malady prevention that might motivate her toward genetic testing, she could be assured that adopting certain diet measures, and so forth, might benefit her as much whether or not a genetic linkage existed.

The situation of genetic testing seems different from the usual medical ethics truth-telling cases in which most persons, at least in North America, would probably say that they would want to know the truth. In most of these cases, (1) the person already knows that something may be seriously wrong, and (2) the harms that may await the person will occur in the relatively near future. Because he knows that something may be seriously wrong, he is apt to worry almost as much if he does not know the facts as if he does know them. Because the harms may occur in the relatively near future, the length of time he can carry out his life normally (if he can deny his fears) may be fairly brief, and the planning advantages of knowing the probable future are far greater. These factors tip the equation so that learning the truth is almost always rational and avoiding the truth is often irrational.

Huntington disease, by contrast, provides an example where not knowing one's genetic structure may be rational. Persons in their early twenties, who know that they have a 50% risk of developing the disease in ten to twenty years vary in their desire to know their genetic status, and it is generally felt that the desire to know and the desire not to know are both rational, although recent studies have shown significant benefits from knowing. However, to have one's genome tested to see if there is a risk of developing just any serious malady prematurely is significantly different. People know that they are at risk for Huntington disease and so not knowing still leaves one with considerable anxiety. However, in the absence of any known risk factors, one would have a relatively small chance of discovering through testing that one had a significantly increased risk of suffering a serious malady. Most of us vaguely appreciate that we are at some small risk of this kind already, and so staying in the dark about this matter would not change our lives appreciably. Most of us can deny—that is, not think about—likelihoods of that sort rather easily. The denial "works" for decades, and for about half or more of us, who die at or near our average life expectancy, it does turn out there was nothing to be worried about all along.

If this analysis is correct, it is very likely that new genetic discoveries will uncover some genetic predispositions, which may or may not count as maladies, that for most persons with average temperaments it is at least as rational not to know about as it is to know about. Thus, although it now seems that those at risk for Huntington disease are psychologically better off knowing than not knowing, this may not be true for genetic conditions about which one has no suspicion. Thus we might have the paradoxical situation in which the increased knowledge created by new genetic discoveries may lead to the view that is not always better to know more, even about one's own genome.

Endnotes

1. See Malady: A new treatment of disease, *Hastings Center Report* 11: 29–37, 1981, by K. Danner Clouser, Charles M. Culver, and Bernard Gert.
2. See Clouser, Culver, and Gert, 1981, for a review and analysis of other definitions.
3. In the Japanese translation of the book *Philosophy in Medicine*, the translator of the chapter on maladies claimed that Japanese did include such a general term. However, on talking with some people who were native speakers of Japanese, we discovered that they would never use this term to refer to a broken arm. Because the translators were physicians, we believe that, like some American physicians, they had mistakenly taken their technical use of the Japanese equivalent of "disease" as representing the ordinary use of that term.
4. For a fuller explanation of volitional disabilities and for a discussion of the distinction between voluntary and intentional actions, see Culver, Charles M. and Gert, Bernard, *Philosophy in Medicine*. New York: Oxford University Press, pp. 109–125, 1982.

8

Morally Relevant Features of Genetic Maladies and Genetic Testing

Like all other maladies, genetic maladies vary greatly in their seriousness. Unlike other maladies, genetic maladies can be discovered long before there are any symptoms. These characteristics of genetic maladies require us to consider what features of genetic maladies we should use in making both private and public policy decisions concerning whether or not to test or screen for the presence of the malady. We call those features that we believe should be used in making these decisions the morally relevant features of genetic maladies. In this chapter, we concentrate on the morally relevant features that determine the seriousness of the malady, which should be used in both private and public policy decisions, but we also discuss other morally relevant features that should be used only in public policy decisions.

Genetic Maladies

Genetic maladies constitute a large and diverse set of medical conditions that share two essential features. First, they are maladies, given the definitions and criteria presented in Chapter 7. Individuals born with a genetic malady may die prematurely, or suffer pain, or be disabled, and so forth, or they may be at increased risk of doing so, as in the case of a hereditary predisposition to develop cancer or heart disease early in life. Second, genetic maladies are caused, at least to some extent, by the inheritance of one or more particular genes, from one or both parents. The current total number of medically recognized human genetic maladies is more than 2,000, although most of these are rare. We expect, however, that in the next ten years this number will probably double, as improvements in genetic technology lead both to the discovery of new, rare maladies and to the identification of a genetic component for already known maladies (e.g., diabetes, muscular dystrophies, mental retardation).

One consequence of this new knowledge will be the tendency to assert the presence of a gene, or genes, that affects or determines some important human trait, such as sexual orientation or intelligence, or that contributes to or causes a malady of unknown etiology, or that predisposes affected individuals to develop a malady such as cancer. These assertions will often be based on inconclusive data that are supported by a strongly held belief or suspicion and that reflect a growing public and scientific sense that DNA and genes are primary determinants of human structure, function, behavior, and affect. There is a long history underlying the "ideology" of genetic determinism, a history that alerts us to the likelihood that the new genetic discoveries would also lead to assertions about the significance of "genetic differences" between people, between groups of people, or between the sexes that are not justified scientifically. Spurious reports of the discovery of a gene for alcoholism, or manic-depression, or criminality will later be disproved or retracted, but these rebuttals or retractions will seldom if ever be given the media coverage provided the original claim. As a consequence, these incorrect assertions and reports will often become accepted uncritically by the public and even by so-called experts.

A second and more immediate consequence of the increased knowledge about the human genome and human genetic maladies will be an increased effort to use that knowledge in order to develop genetic tests. We have recently seen the commercialization of genetic tests for Huntington disease, cystic fibrosis, fragile-X syndrome, and muscular dystrophy, and we believe that many other molecular tests for genetic maladies are in the development stage. It is plausible to imagine that within a decade physicians will have several dozen tests available that can be administered on a large scale, for screening newborns or at-risk populations. They will also have tests to be provided on a small scale: (1) to inform concerned couples about the potential risk of their having affected children, (2) for prenatal diagnosis when both members of the couple are known or highly suspected of carrying a deleterious gene, and (3) to determine whether adults will develop late-onset disorders such as Huntington disease or breast cancer.

The availability and widespread application of such genetic tests will raise two sets of new problems. The first involves deciding which of the many genetic maladies should be subject to testing. Once the tests that are developed become accepted as "standard medical practice," physicians will have a duty to inform clients of their availability and to explain both the risks and possible outcomes of a decision to have the genetic test *and* the consequences of not having the test. Because the client/patient must have all the information necessary to make a rational choice, the burden for providing the necessary information will fall on medical professionals. Because there are clear legal implications associated with widespread genetic testing—for example, what counts as adequate information for valid consent to or refusal of testing—it will become important to decide which maladies ought to be included for testing and what information the physician/counselor should provide to the patient/client.

Second, major resource allocation issues will be created particularly in relation to providing the cost of testing and to providing the increased level of genetic and psychological counseling that will be required for the large numbers of people being tested. There is already concern about the limited number of genetic counselor training programs available to supply the technical expertise that will be needed to carry out and evaluate the tests. It is likely, at least initially, that genetic testing will be expensive both to society in general and to individuals and families who request testing. Few if any insurance companies will cover this expense—except if it is at their request and with specific conditions—so that most people will be ineligible for compensation by health insurance companies. So, for purely economic reasons, there will certainly develop some limit to the resources available for genetic testing and, consequently, there will emerge some governmental or insurance industry policies that address genetic testing.

There are only two possible overall strategies for the development of policies governing genetic testing. One is to decide on a case-by-case basis whether a new genetic test should become publicly available and thus be included in the class of tests that are considered "standard medical practice." Any test included in this category would be readily available and invariably recommended when the family history or other medical circumstances deem it appropriate—for example, affected sibs or relatives, maternal age, previous miscarriage, or ethnicity. The second approach is to establish, initially, some explicit set of standards or criteria that must be met before a new genetic test becomes considered "standard medical practice" and is administered broadly.

Our view is that a case-by-case approach, if it is not to be completely idiosyncratic, must rely on these standards and criteria, at least implicitly. We believe that proposing an explicit, systematic set of public standards and criteria for determining which tests should become "standard medical practice" is more likely to minimize inconsistencies and prevent decisions based on unrecognized biases. We realize that an explicit statement of standards may be premature and that new facts or considerations could emerge that would change our views concerning certain criteria. Thus, we think it important to emphasize that the standards are provisional, open for discussion, and certainly subject to change as we gain increased knowledge and understanding. In terms of public policy the final standards necessarily will involve some balancing of costs, risks, and benefits, just as setting speed limits does; and, as with speed limits, there will be a range of morally acceptable policies. In terms of deciding which genetic maladies deserve testing, in private contexts, a similar systematic form of analysis will also be needed.

Our effort, in this chapter, is to anticipate the two kinds of problems that will arise—namely, which maladies should be included for testing and how to deal with the resource issue, and develop a rational approach, proactively. In doing so, we begin by defining and justifying what we term the "morally relevant features" of genetic maladies. These are the important aspects of a genetic malady that affect any moral decision we make about it. This decision may be whether or not the specific testing or screening procedure should become "standard medical practice," which is a public policy issue, or whether to continue a pregnancy involving an affected fetus, which is a private counselor/physician to client/patient issue. As a start, we examine what is generally termed the "seriousness" of a malady and ask what aspects enter into the assessment. We choose "seriousness" because it is a term that is easily understood and widely used. The features of a genetic malady that determine what we call its seriousness are morally relevant to all decisions, private and public, that are made about it. Other morally relevant features, such as the cost of testing, are relevant only to public decisions—for example, whether widespread testing should be adopted.

Seriousness of a Malady

In discussing a patient's condition, we commonly refer to the "seriousness" of the medical disorder. Melanomas or occluded coronary arteries are considered to be very serious maladies, whereas chronic prostatitis or a simple fracture of the tibia are said to be less serious. In the case of pattern baldness, a genetically determined and sex-limited trait, most people would not consider it serious and many would even question whether to call it a malady at all. The features of a malady that affect our assessment of its seriousness are among the most important aspects of genetic maladies. Serious genetic maladies should justify concern and lead to the widespread application of genetic screening and testing, and such genetic maladies will present the greatest challenge to the physician/counselor–patient/client relation in terms of providing information and support. Because we are interested primarily in providing a moral justification for the development of genetic testing and screening, we will try to separate out those "morally relevant" features of genetic maladies from economic or resource allocation issues that may influence public policy. We are also interested in determining whether or not genetic maladies are different from nongenetic maladies, in any important way, in terms of the kinds of features that affect our assessment of seriousness. The following features could contribute significantly to our assessment of the seriousness of a malady, regardless of its origin.

Mortality Death is generally the most serious outcome of any malady, genetic or nongenetic, and so the seriousness of a disorder is *prima facie* related to the degree to which it threatens the patient's life. In many cases, mortality is age related. Pneumonia is a nongenetic malady that is potentially life threatening, but the mortality associated with pneumonia is highest in the very young and in the very old and is relatively low at other ages, all other things being equal. Age of onset and age of death in this and other cases become major factors in considering the "weight" given mortality. We will talk about age of onset and age of death in a separate section.

Pain and Disability Physical or psychological pain or disability is the second most commonly recognized morally relevant feature of a malady. No rational being chooses to experience pain for no reason at all, and physicians now recognize the importance of minimizing the severe pain associated with terminal maladies such as cancer through the use of opiates or other analgesics. An individual with a toothache, an earache, or a broken ankle probably experiences severe pain, but these maladies are not generally considered serious because, in spite of the intensity, the *duration* of pain experienced is relatively short. Maladies involving little pain, such as multiple sclerosis, are still considered exceedingly serious because of the associated mortality and because they lead to severe disabilities. Maladies in which there is severe, chronic pain may often be so undesirable that the afflicted individual may prefer to die. The same holds for an acquired disability such as blindness or total paralysis, which can be regarded by the person affected as being as serious as a very high risk of premature death. It is important to distinguish between the loss of previous ability, as in the cases considered here, and the acquisition of a disability at birth, which may be perceived quite differently by the affected person.

Loss of Freedom or Pleasure In some cases, a malady may be considered to be very serious even when there are few or no symptoms. At present, HIV seropositivity is a good example of a malady that is currently symptomless but will eventually lead to death, pain, and disability. There is no question that death, chronic pain, and permanent disability are generally considered more serious than long-term loss of freedom or pleasure. Loss of freedom (as in allergies) and loss of pleasure (as in some sexual disorders) are generally considered much less serious. Finally, there are environmental and social contingencies that affect the level of seriousness associated with a disability. Our society will increasingly be encouraged to provide social accommodation to disabilities such as blindness and confinement to a wheel chair. Our perception of their seriousness will, as a consequence, diminish.

Treatment Another important aspect of seriousness is whether or not a malady can be treated successfully. Treatment in the sense used here refers to a physician's ability to reverse the progress of a malady. Many treatments may slow or even halt the progress of a malady without leading to recovery. For many terminal maladies, "treatment" may simply involve relieving the symptoms, such as pain or disability, without effecting any real "cure." But, for treatment to count as an important aspect of seriousness, the notion of "reversing the progress" or halting the progress early on are critical. AZT treatment, for example, can temporarily slow the progress of AIDS, but AIDS is a very serious disorder because it inevitably leads to death. The treatability of a malady has the most effect on the seriousness of a malady when the untreated malady leads to premature death, severe chronic pain, or significant permanent disability. Chronic prostatitis, in contrast, may often be untreatable but is not considered to be a very serious malady because, even untreated, it seldom leads to death, severe pain, or disability. It seems as if mortality, chronic pain, and significant permanent disability combine with other aspects such as treatability in determining seriousness. A malady that leads to death is much more "serious" if it is untreatable (e.g., AIDS) than if treatments exist (e.g., diabetes or pneumonia). Needless to say, the successful treatment or cure for a disorder could appear at any time, thereby changing the status of that disorder in terms of "seriousness." Infectious diseases caused by bacteria were, until the 1940s, considered to be very serious maladies, but with the application of antibiotic treatment most infections no longer have or deserve that status.

Conclusions about Seriousness There are probably a significant number of people who hold that premature death is the worst outcome of a malady. For many others, there are fates worse than death. But, if we were to take a poll among members of that group and ask what those fates are, we would likely uncover a significant similarity in their responses. For some, extreme, chronic pain is worse than death; for others, significant permanent disabilities (say, quadriplegia) might top the list. But it is safe to say that, even among this group, premature death ranks very close. So our assessment of the seriousness of a malady appears to correlate with the mortality, severe chronic pain, and significant permanent disability associated with the malady. The level of seriousness of a malady is often a direct consequence of its treatability or the ability to prevent a malady from occurring by, say, immunization, prophylactic treatment, or simple life-style adjustment. So the status of a malady, in terms of the concern it arouses, can change rapidly as a consequence of some medical discovery that provides a means of preventing death, pain, or disability and restoring normal health. We have no reason to believe that genetic disorders differ from nongenetic disorders in terms of these features or aspects of seriousness.

Unique Aspects of Genetic Maladies

There are, however, several aspects of genetic maladies that do distinguish them from many nongenetic maladies and that deserve some attention especially in terms of counseling. The first has to do with the "age of onset," and the second involves the so-called "range of variation."

Age of Onset Many nongenetic maladies can occur in people at any age. Infectious diseases, malignancies, and accidental injuries, for example, may affect either sex at any age, although malignancies generally occur in late adulthood. Genetic maladies, in contrast, are often characterized by relatively specific times of onset. Chromosomally based disorders (e.g., triploidy) generally produce their lethal effects during the first trimester of fetal life, and death may even occur before uterine implantation. The fraction of conceptions that terminate because of such genetically based developmental maladies could be very large indeed. The onset of anencephaly, which may be to some extent genetically determined, occurs in the second trimester, although mortality occurs at birth. Inborn errors of metabolism often manifest themselves in newborns, and maladies such as Tay-Sachs or Lesch-Nyhan lead to death in early childhood. In spinal muscular atrophy, death occurs in adolescence, and, although Duchenne muscular dystrophy, cystic fibrosis, and juvenile diabetes are not necessarily lethal, their effects become manifest primarily in childhood, adolescence, or young adulthood. In several cases, such as Huntington disease, adult polycystic kidney disease, and hereditary Alzheimer's disease, the onset of the malady is often later in life in the postreproductive years. We believe that the specific age of onset affects our assessment of the seriousness of genetic maladies and becomes an important morally relevant feature. Maladies that affect infants, children, and young adults in most cases arouse the greatest concern. Some argue that this assessment is discriminatory but that interpretation is wrong, because unlike ethnic discrimination where only a few are affected, in the case of age of onset all people age and all people count. To illustrate the effect of age of onset, consider two hypothetical cases, in the context of new genetic discoveries.

Imagine that through new genetic investigations scientists discover two new genes and two new genetic maladies. In the first case, the mutation of some gene invariably causes its carrier to undergo cardiac arrest at the age of ninety. In the second case, a mutation of the gene produces death at the four-cell stage of embryogenesis, well before uterine implantation occurs. Both maladies have existed for millennia but have gone undetected until DNA technology uncovered the presence and location of the relevant gene. The first malady was not recognized because it closely resembles the general process of senescence, and the latter was missed because there was no evidence that a conception had occurred. In fact, we later come to learn that many of the families who are unsuccessfully treated at fertility clinics carry a mutant form of the relevant gene. In both cases, the defective gene is directly responsible for death and would seem to fit into the class of very serious maladies. But, if we compare them with, say, Lesch-Nyhan syndrome or Tay-Sachs, we sense an ambiguity.

In the first case, our qualification is clear. Most people are already dead by age ninety, and so dying of a heart attack at that advanced age seems neither unnatural nor abnormal. In fact, the death certificates for patients later shown to carry that gene list the death as due to "natural causes" without specifying any malady. If medical technology discovered a means of extending youth to age ninety by delaying the process of aging and senescence, death resulting from a heart attack at age ninety would be a serious malady, whether or not it is genetically based. Needless to say, on the day that one dies of a malady, that malady is quite serious regardless of age or etiology. Rational people would not choose death for no reason, and rational people who are not suffering from severe chronic pain or significant permanent disability spend lots of money for tests and treatments when a life-threatening malady appears, regardless of their age. The degree of seriousness of a disorder, in terms of age of onset, is additionally related to life expectancy. To that extent, it is a normative feature containing significant cultural, ethnic, and national variation.

In the second case, in which development terminates at the four-cell stage, one could argue that, because the four-cell stage embryo is a potential moral agent that is alive and can therefore die, this genetic malady is serious because it leads to such a premature death. However, rational people can legitimately disagree about how serious this malady is. In this case, the termination of embryonic development occurs prior to the stage at which many think it appropriate to become seriously concerned with whether an organism lives or dies. However, even if we accept that the embryo is not sufficiently developed to be an object of serious concern on its own, we still might regard the malady as serious *for the couple* who are unable to have a child. That is, we might regard the gene primarily as causing a fertility problem rather than being a malady leading to death. This case shows that, when we are dealing with the death of someone who is not even a sentient being, we tend to be more concerned with the effect of that death on the people affected by the death than by the death itself. The more we regard an organism as an already existing person, the more seriously we regard a malady that affects that organism. (See Chapter 9 for more discussion of the concept of a person.)

Range of Variation A second feature of many genetic maladies is their variability. Geneticists have known for some time that there is frequently no one-to-one association between the presence of a specific gene or genotype and the expression of that gene or genotype at the level of the phenotype. Genetic terminology has been developed to communicate that uncertainty, and geneticists often talk about genes with incomplete penetrance—that is, not everybody with the same genotype expresses the trait in the same way or to the same extent. Cystic fibrosis and Down syndrome may have 100% penetrance—that is, all carrying the defective genes show some phenotypic abnormality—but the expressivity of those maladies is variable and the abnormality may manifest itself in a range of severity among individuals with the same genotype. Factors that influence the range of variation, penetrance, and expressivity fall into four categories. The first is the so-called genetic background—that is, all the other genes that alone would not produce or lead to a manifest phenotypic malady but that may interact with or modify the expression of the malady-causing genes to enhance or reduce their expressivity. We know a great deal about modifier genes in so-called model genetic systems, such as *Drosophila* and the mouse, but are only at the edge of understanding the kind and number of modifier genes in human beings.

Environmental variation is another important source of the incomplete penetrance and variable expressivity for many genetic maladies. The effect of dietary protein on adult size has been presented as the major factor accounting for the larger size of American-born Japanese. The pattern of pigmentation in Siamese cats is entirely due to ambient temperature. In armadillos, all members of a litter are genetically identical, each member having been formed from a single mitotic cell produced by post-fertilization cell division. Yet the range of litter-mate size may vary enormously. This size variation arises as a consequence of the site in the uterus where the embryo implants, because implantation sites vary in terms of maternal blood supply. Implantation at favorable sites will lead to a large and vascularized placenta; implantation at a marginal site leads to a smaller, less enriched placenta. Here, then, perinatal environmental variation alone can lead to phenotypic variation among genetically identical sibs. Analogous situations must exist in human beings. Postnatal nutrition also is an important factor in generating phenotypic variations. A classic case is the Shetland pony, whose small size in the Shetland Islands is due to the lack of an essential nutrient in the diet. Shetland ponies brought to the United States and raised on Kentucky Blue Grass grow into adults larger than their parents. But it takes several generations for the transported ponies to achieve full size; that is, in spite of a complete diet, full-sized adults do not form in the first generation. This is because the maternal uterus in a small Shetland pony cannot support the growth of a full-sized colt, even if her diet and that of her offspring are complete, and reduced birth size limits future adult size.

Third, there are stochastic or nondeterministic processes that occur during development and lead to a range in phenotypic variation. Consider, for example, the dog breed Dalmatian. Although we can be certain that the offspring of two spotted Dalmatians will be spotted, we cannot predict where on the coat black spots will appear. The asymmetry of left-right fingerprints in human beings, even identical twins, can be explained by chance events alone. It is quite possible that many of the interesting differences between people are produced by such developmental "noise." Fourth and finally, there may be multiple alleles—that is, several alternative mutant forms of a gene. Although each mutation leads to some degree of phenotypic abnormality, the alleles differ from one another in the extent to which the abnormality is expressed. Cystic fibrosis, as we will see, is a good example. Before we move to real world situations, there are two additional points that need to be made concerning the range of phenotypic variation expressed by a single gene or genotype, as it pertains to the variable expression of genetic maladies.

The Norm of Reaction Phenotypes represent the expression of genotypes, in the context of a particular environment, in the presence of developmental noise. However, the correlation between phenotype and genotype does not necessarily remain constant over the full range of potential environments. The range of phenotypes expressed by a genotype, over a range of environments, is known as its norm of reaction. The shape of the curve, however, characterizes only that particular genotype. The norm of reaction for another genotype may be different and even show a distinctly different shape. It is likely that, if in human beings there are different genotypes that can predispose one to develop hypertension, the norms of reaction, as a function of, say, dietary sodium or of stress, will be different for each.

The majority of human maladies that will be shown to have a genetic basis in the future will likely be multifactorial or polygenic. Here the disorder (hypertension, heart disease, tumor formation) results from the expression and interaction of several or even many different genes; that is, it is polygenic. The range of expression for polygenic traits is usually affected by variation in the environment and is thus multifactorial. The term heritability has been coined to reflect that complexity. Heritability is a value, which ranges from 0 to 1, that refers to the proportion of total phenotypic variation in a population or group based on underlying genetic variation. Obviously, genetic maladies with high values of heritability will, because of their predictability, be much easier to deal with, in terms of genetic counseling, than will maladies with low values of heritability. It is likely that the majority of maladies found to have an identifiable genetic component will be polygenic, will have complex norms of reaction, and will display low values of heritability. The inability to make direct associations between genotype and phenotype becomes a novel and serious problem not only in studying such maladies scientifically, but, as we will see, also in dealing with these maladies at the level of public policy, for screening and testing and for medical genetic counseling. Of particular concern will be unsubstantiated claims that behavioral and affective maladies show high values of heritability. Similar mistaken claims in the past have been used to justify onerous programs of eugenics.

Cryptic Variation There are many normal traits and many maladies that have a clear, unambiguous genetic basis. In fact, the process used by geneticists to determine the nature and map location of genes, including those that produce maladies, is to follow the familial pattern of inheritance over several generations. Genes affecting morphology (size, shape, color, digit number, and so forth) and metabolism (enzyme deficiencies) are the simplest to identify because each person can be assigned a discreet phenotypic value. Genetic variation affecting traits that are termed physiological (blood pressure, renal output, visual acuity) or behavioral (moves away from light, learns a maze quickly, and so forth) are more difficult to analyze systematically because the method for assigning a phenotypic value can be complicated or even ambiguous. These traits also tend to have a polygenic basis, which further complicates genetic analysis, for reasons already discussed.

The most difficult of all traits to identify and study genetically, however, involves what we refer to as cryptic variation. That is, variation present in families or in populations that is not obvious until the individuals or population has been challenged or perturbed. Traits that can be measured only in this way include the ability to adjust or to even survive stress (heat, cold, radiation, high altitude), aspects of performance (memorizing lists, lifting a heavy weight), and predisposition (the tendency to develop some physical or mental malady in response to one or several specific factors in the environment). Although genes that predispose people to develop specific maladies such as cancer, heart disease, or mental illness are of great interest and concern, both in the medical and in the social sense, they are difficult and often impossible to analyze scientifically. As a consequence, a great deal of social controversy will surround claims that identify specific genes that predispose individuals to maladies such as arteriosclerosis, cancer, schizophrenia, or alcoholism.

Private and Public Issues Related to Testing

Perhaps the most important role of the physician/counselor in medical genetics is to provide relevant, balanced, and accurate information to the patient/client about the genetic malady of concern. In principal, this permits the patient/client an opportunity to make rational and informed decisions, if options exist, or to plan appropriately if there are no choices. By examining the nature and significance of the morally relevant features of genetic maladies, we hope to provide the physician/counselor with a framework for describing and classifying all genetic maladies in a way that will be most helpful to the patient/client. One outcome of our analysis, then, could be the construction of a profile, perhaps in table form, for each genetic malady, including new ones, and a classification scheme that groups genetic maladies with similar profiles. Such a scheme would be a useful tool for the physician/counselor in providing the patient/client concerned with a new or rare genetic malady with an account of the morally relevant features of that malady and perhaps even a set of reference genetic maladies with which the patient/client may be more familiar and about which there is more information.

This system could easily become automated into computer software or interactive videos, allowing the patient/client to learn about her genetic malady, or that which her fetus may have, slowly, deliberately, and with the option of returning to issues or features presented earlier. This uniform and unbiased method of providing information does not require the expenditure of the valuable and limited physician/counselor time. It is important to point out that this system of feature profiles is not intended to rank, or even capable of ranking, genetic maladies. Different people will always come up with different rankings according to their own values and beliefs; and it is important for the patient/client to know that rational people rank maladies differently. Moreover, the calculus of ranking requires quantifiability, and we know that many of the morally relevant features discussed, such as pain and disability, cannot be scaled in an unambiguous way. Because of the strong tendency to conclude mistakenly that there is only one correct ranking we would, in fact, strongly argue against the attempt to simplify the set of morally relevant features by giving them numerical values, even from the most prudential and benevolent motives.

There are several morally relevant features or aspects of genetic maladies that we have not yet discussed. These include the frequency of occurrence and the costs of treatment care and testing. Although these issues do come up in private counseling sessions, they are seldom critical factors in the decision-making process used by patient/clients to make personal choices. However, these features do become important in considerations of public policy, such as the decision to include a new genetic test as standard medical practice or to establish a large-scale, carrier screening program. Because our intent is to look at both the issues of medical standards of practice and the rationing of genetic resources, it is appropriate to consider these public issues next.

Cost of Care The financial burden of a genetic disorder, to the patient and his or her family, can be enormous and as a consequence may contribute to the seriousness of the disorder. In those cases for which insurance covers the cost of treatment or maintenance, the cost of treatment will be an important feature of a disorder in the social policy sense. It may cost millions of dollars to successfully provide treatment to the premature babies born weighing less than a pound but only a tiny fraction of that amount to treat a serious case of pneumonia. We recognize that physicians are bound by professional duty to save lives, unequivocally, and in that sense the cost of treatment or care carries little weight in making personal moral decisions. However, on occasion various public policies do set cost constraining limits on what a physician may do, thus relieving the physician from making decisions in her private practice that would create tension between the ethical practice of medicine and the realities of resource limitations. The likelihood of the one-pound infant surviving and of subsequently developing into a normal child is quite low, whereas antibiotic treatment of pneumonia is nearly always successful in curing children and adults. So, although we would prefer to provide full medical attention to both conditions, in the public policy realm it makes more sense to concentrate resources for the treatment of pneumonia than to invest in more technically sophisticated and costly very low birth weight infant life-support systems.

Incidence As stated initially, most genetic disorders are rare, appearing at incidences in the range of $1/10^4$ to $1/10^6$. There are exceptions of course. Among those genetic disorders with a higher rate of incidence are (1) Down syndrome among children of women who become pregnant after age forty-five, (2) cystic fibrosis among Caucasians, (3) sickle cell anemia among blacks, and (4) thalassemia in populations derived from the Mediterranean area. The importance of incidence as a morally relevant feature involved in public policy is derived from risk-benefit or cost-benefit analysis. Genetic testing is not prudent in those cases in which the risk of doing a test, to either the fetus or mother, is greater than the malady's incidence or if the cost of screening a population in order to prevent a malady greatly exceeds the cost of caring for or treating individuals born with the malady when no screening is done. If screening newborns for phenylketonuria (PKU) produced far more deaths from infection than were prevented by detecting and then treating PKU, it would make no sense to screen. It is clear that the morally relevant feature here is "how many" will be affected in a deleterious way, not "who gets" the malady. Unfortunately, a trend is evident for interest groups representing a particular hereditary malady to garner publicity, resources, and support at a level that is out of proportion to the actual incidence of the disorder.

Types of Genetic Tests

The most common type of genetic test is known as prenatal diagnosis, and the most common malady tested for is Down syndrome. Because prenatal tests are used to determine whether a high-risk fetus actually has or will develop the genetic malady, the outcome of a positive result is usually abortion. In the minority of cases in which abortion is not an option, a positive result may be important for financial, medical, and psychological preparation. Because prenatal testing involves some risk to both the mother and fetus, prenatal tests are done only for serious maladies and only when there is a significant likelihood that the fetus is affected. In the majority of cases in which prenatal testing is used, the malady involves premature death or significant permanent disability or both, is not treatable, and is highly penetrant and expressive. In a few cases, such as PKU, prenatal testing permits early intervention (surgery or dietary control) and successful treatment.

Newborn testing is a broad based and often, but not always, legally mandated procedure used to identify infants affected with a genetic malady that is treatable. Nutritional and metabolic maladies, such as galactosemia and tyrosinosis, are included here. Because the majority of newborn tests are cheap and risk free, it is reasonable, both in the medical sense and in terms of cost-benefit analysis, to administer tests broadly. For the same reasons, unless the malady is trivial, it makes no difference how serious it is—only that there is available treatment. When there is no treatment, newborn testing is generally not recommended even when the age of onset is in childhood. When the age of onset is not until adulthood, and no treatment or prevention is available during childhood, newborn testing may even be prohibited.

The third type of test is genetic screening. Adult screening is often targeted at populations or ethnic groups known to show a high incidence of some genetic malady. The most well known examples include Tay-

Sachs disease among Ashkenazi Jews, the thalassemias among Greek and Italian populations, sickle cell anemia in populations of African ancestry, and cystic fibrosis in Caucasian populations. Screening is used to identify carriers and couples at risk of having affected children. The results of screening are important in reproductive counseling and family planning. At-risk couples, in which both are carriers, once identified are counseled to consider prenatal testing, or adoption. Genetic screening programs deal, almost exclusively, with genetic maladies that are serious in terms of premature death, pain, and disability; and for which cost-benefit (related to incidence and cost of care) and risk-benefit (related to fetal mortality resulting from the test) are quite favorable. Screening programs are almost always voluntary and supported by government or private agencies.

The Ethics of Genetic Testing

The morally relevant features of a genetic malady that are included in the concept of seriousness are important in two different contexts. First, they are among the topics discussed extensively when the physician/counselor is talking to the patient/client. In this *private* context, they represent the features of a genetic malady that arouse both medical and personal concern such as premature death, pain, disability, age of onset, and so forth. They are the features that strongly influence personal decisions and choices, such as whether to marry, whether to have children or to adopt, and whether to terminate a pregnancy. Decisions related to the development and application of genetic tests, at the private level of analysis, are almost exclusively based on these morally relevant features.

Second, these same morally relevant features contribute to, but do not determine, decisions related to public policy, such as whether to introduce a specific test as a standard of practice or to establish a large-scale, adult genetic screening program. However, decisions in the area of public policy depend on other features that are not relevant to the private context, such as the incidence of occurrence or the cost of testing, screening, or treating. Many would claim that moral considerations alone rarely drive anything on the public policy level. Admittedly, programs such as genetic testing are driven more by economic, legal, and political considerations, but each of these considerations is also subject to moral evaluation. We are now concerned solely with the morally relevant features that determine the seriousness of the malady and how they might effect whether we should have some new prenatal genetic test or introduce some newborn or adult screening program. If we lived in a world where people were guided by these morally relevant features, how would public policy issues related to genetic testing be decided?

For newborn testing, treatability seems to be the most important feature. It would be irrational to test newborns or children for an untreatable genetic malady, unless the test was critical for making a correct diagnosis or if the results of the diagnostic test are important to the family in terms of preparing financially for an illness that will occur later in childhood. It makes no sense, and even may be morally prohibited, to test newborns or children for late-onset disorders, such as Huntington disease or familial Alzheimer's disease, maladies for which no treatment or cure exists. Adult testing for late-onset disorders raises no moral concern, as long as testing is voluntary, confidential, accurate, and risk free.

Because prenatal genetic testing almost always implies the option of abortion, the morally relevant feature of greatest importance here is the overall seriousness of the malady. Genetic tests for sex, eye color, or stature, even if available, would raise serious ethical problems. None of these traits are maladies; so, if a justification is needed for abortion, a fetus's sex, eye color, or future size normally would not provide such a justification. Given that the demand for genetic testing will outstrip the supply for the foreseeable future, we do not think that prenatal genetic tests for nonmalady genetic conditions should be developed or offered. Indeed, it may even be inappropriate to develop or offer genetic tests for trivial or mild genetic maladies.

For adult carrier screening of at-risk couples or populations, the overall seriousness of the malady again seems the most crucial feature. In fact, both adult carrier screening and prenatal screening seem very similar in terms of justifying the use of scarce resources. Couples or individuals found to be at risk by adult screening would logically choose to consider prenatal screening an option when a pregnancy occurred. Needless to say, adult screening would not be justified in cases in which the risk of harm produced by taking the test was greater than the risk (the rate) of having an affected child.

Genetic Tests Available

It is now appropriate to examine several cases in which a newly discovered gene has led to the development of a molecular test capable of identifying the genetic status of an individual—that is, of distinguishing whether the individual being tested is free of the defective allele, is a carrier, or has the genotype associated with the disorder. As mentioned in Chapter 5 concerning dilemmas involved in counseling for Huntington disease, there is now a direct test for the *HD* gene.

Cystic Fibrosis Cystic fibrosis (CF) is a common genetic disorder in the Caucasian population (incidence of 1/2500 births) and is caused by the absence of a normal allele (see Appendix, Figure 6) for a gene required for the production of a protein involved with the transport of chloride ions across the cell membrane. Although the defective gene is present in every cell, the disease state manifests itself in the lung, sweat glands, and pancreas. Cystic fibrosis has been considered an early-onset, lethal genetic disorder, and in the past affected children rarely survived to adolescence. It is known to show a wide range of severity, which we now know is partly based on the presence of a large number of different mutations, each producing its own characteristic range and level of severity. Treatment had been limited to chest "thumping," a technique used to dislodge the viscous mucous produced by CF children, and antibiotic aerosols to reduce the likelihood of lung infection. Recently, however, aggressive treatment of CF children has extended their life-span and it is not uncommon to find CF patients in their late teens and twenties. Progress in the development of new treatments, such as the DNAse aerosols developed by Genentech, promise to further extend the survival of CF patients.

The use of somatic cell gene therapy, in which an infectious viral vector is used to transfer a normal copy of the CF gene into the cells of the bronchial epithelium, although currently in the early stages of development, will likely eventually lead to a real cure for cystic fibrosis. For cystic fibrosis, it becomes crystal clear that what used to be a very serious genetic disorder (early death, pain, disability, the absence of a cure) will probably become significantly less "serious" in the near future. In this case, the change in status is based both on more effective treatments, which reduce the progress of the disorder and prevent the lethal effect of the mutation, and on the use of gene therapy to cure the disease. One outcome of this change in status is that the criteria that must be met in order for CF genetic testing to be adopted broadly in a national screening program will become more stringent. So, in this case, for which current screening tests can detect the presence of only the more common mutant forms of the CF gene (about 70%–90%, depending on the ethnic group tested), the decision *not* to screen broadly for the purpose of abortion would become increasingly more justifiable as the ability to treat or cure cystic fibrosis increased. The caveat is that, if there is a treatment for a disorder such as cystic fibrosis and if it is effective when begun prenatally or at birth, then testing with the goal of determining whether or not to begin treatment would be easily justified.

Myotonic Muscular Dystrophy The genetic alteration responsible for myotonic muscular dystrophy (MMD) has been localized to chromosome 19. The structural defect associated with this disorder is remarkable in that it may grow larger with each generation, and this growth is associated with an increased expressivity; that is, the severity of the disorder worsens over generation time. The MMD gene contains an internal coding sequence of three nucleotides, CTG, from five to twenty tandem copies of which are found in normal individuals. People having from fifty to one hundred CTG copies show mild symptoms, and individuals with severe symptoms may contain genes with thousands of copies. This unexpected genetic situation has actually been found to be responsible for other genetic disorders: the fragile-X syndrome, which causes mental retardation; androgen receptor deficiency, which leads to abnormal gender phenotypes; Huntington disease; and one form of colorectal cancer. Myotonic muscular dystrophy is a late-onset disorder, first appearing in adolescence and young adulthood. The other genetically determined form of muscular dystrophy is the Duchenne type, which has a very early age of onset. Furthermore, MMD is an autosomal dominant disease (see Appendix, Figure 5), whereas the Duchenne type is a recessive X-linked disorder (see Appendix, Figure 7).

Myotonic muscular dystrophy causes muscular spasms, weakness, and atrophy of voluntary muscles. Often, patients experience myotonia, a difficulty in relaxing muscles. The disease, which was first described in the early 1900s, shows, as discussed, highly variable expressivity even among the affected members of a single family. It has become clear that there is a progressive and apparently irreversible increase in CTG copy number over generation time. To our knowledge, it is not yet known whether the expansion of the gene is based on some error in DNA replication or is a consequence of some other process, such as unequal crossing-over during recombination. Nor is it known whether the expansion can occur in somatic cells as well as in germ-line cells. In terms of screening, a unique and perplexing feature of the disorder is that not only can a couple be found to be at risk of having a child with MMD, but in a presymptomatic couple it will be possible to predict that they will have unaffected children but that their grandchildren will be affected. In some sense, then, parents can be involved directly in making reproductive choices that affect family members two or three generations away. To make it more complex, there is now evidence that expanded sequences can contract over generation time, so that, if a parental chromosome carries a presymptomatic but high copy number, the chromosome transmitted to the child can carry more, fewer, or the same number of triplet copies.

Conclusions

Couples confronted with the suspicion or with the knowledge that they are at risk of conceiving a child affected with some genetic disorder generally ask their physicians or genetic counselors the same set of questions. How serious is the disorder? Will the baby die from it and, if so, at what age? Will the disorder produce symptoms that are painful or that are disabling? Will the infant or child's activities be limited in any way? When will the symptoms first appear? How widespread is the disorder? Is there a cure, treatment, or way of preventing the disorder? And, perhaps most importantly, is there a "good" (that is, safe, valid, and reliable) test that can tell us if we are carriers who are at risk for having a conceptus with the disorder or if the conceptus has inherited the disorder? From answers to these questions, the couple acquires the relevant information that will later influence or determine their choices and decisions about marriage, about having children naturally or by adoption, about prenatal diagnosis, or about aborting an affected fetus. Because many of these are moral choices, these features of the disorder that are important in making those choices become the morally relevant ones.

Not surprisingly, the most important morally relevant features are whether the disorder is fatal or is associated with severe and permanent pain and disability. The emotional and psychological devastation to a family caused by the loss of an infant or young child, or his or her inability to develop normally, is immeasurable. That is why the development of valid and reliable molecular tests to detect early-onset, unconditionally lethal genetic disorders, such as Tay-Sachs disease, ADA deficiency, or Lesch-Nyhan syndrome, will rank as a first-order priority. Lethal genetic disorders may become treatable postnatally, however, thus diminishing their seriousness and relieving the urgency for population screening and prenatal diagnosis. A good example is cystic fibrosis, which has already become treatable to the point that affected children may now look forward to living into adulthood and perhaps having their own children. The prospect of curing cystic fibrosis by employing somatic cell gene therapy postnatally will further diminish the level of medical and family concern associated with this disorder. The exception is that in those cases, such as phenylketonuria, in which curative treatment must begin at birth, there is again an urgency in developing sensitive, cheap, and reliable tests. Other examples, like cystic fibrosis, that fit in the category are hemophilia and Duchenne type muscular dystrophy, where genetic biotechnology or muscle cell implantation have developed into potentially powerful treatments and cures.

Untreatable, unconditionally lethal genetic disorders that show late onset, such as Alzheimer's disease or Huntington disease, may be of questionable moral interest in terms of developing genetic screening or prenatal diagnosis simply because of their late onset. Although these disorders are devastating when the symptoms appear, affected individuals have often led productive and happy lives, and so it would not be irrational for a couple to have a child known to carry the defective gene. The possibility that an effective treatment or even cure will be discovered in the next thirty to forty years preceding onset provides further incentive for parents of affected fetuses not to choose abortion. Again, however, when effective treatments or cures for late-onset, currently unconditionally lethal disorders must start at birth or very soon thereafter, the development of valid and reliable genetic tests for screening and prenatal diagnosis does acquire a high priority.

The most difficult group of genetic disorders to assess, in terms of establishing a rationale for the development of genetic testing, include genetic predispositions and disorders based on the inheritance of specific combinations of alleles at several different genes (polygenic traits). In both cases, there is a high degree of prognostic uncertainty, because of incomplete penetrance, variable expressivity, and the important but often unknown influence of particular environmental factors. Whether, and the extent to which, a genetic predisposition to develop heart disease, hypertension, cancer, or some occupational disorder actually materializes is currently unpredictable. Worse still, the stigma associated with being identified as "genetically predisposed" can often be more destructive to an individual than the disorder itself. Because of the uncertainties involved in (1) identifying a predisposed genotype and (2) in then predicting the probability or likelihood and extent to which the disorder will develop, it appears advisable to ignore such disorders entirely as far as genetic testing is concerned, except for completely voluntary adult testing.

The recent discovery of genetic disorders based on "expandable" genes, such as myotonic muscular dystrophy, colon cancer, Huntington disease, and fragile-X syndrome, illustrates the kind of unexpected complexity that is generated by new knowledge produced by the new genetic discoveries. Genetic testing that allows a couple to learn that their potential offspring will likely be normal but that their grandchildren will be affected raises novel moral dilemmas. Equally perplexing are the implications of studies in which it was found that the process of expansion can be reversed over generation time. We would emphasize, finally, that, in exploring the shape and content of the morally relevant features that constitute the seriousness of the malady, we have ignored discussing such issues as the cost of testing and the scientific fine points that affect the validity and reliability of test results. The scientific issues of genetic screening and diagnosis, though very important, cannot be discussed in this chapter. Insofar as the issue of cost is concerned, it only stands to reason that the same principles of cost-benefit analysis used to establish the State of Oregon's policy on medical coverage would apply here. If the cost of carrying out a single test to determine whether someone has an untreatable disorder, such as Huntington disease, is equivalent to that of carrying out one hundred prenatal diagnoses for Tay-Sachs, it is only rational to opt for the latter.

One of the most controversial issues that will come up, with regard to the ethics of genetic testing, concerns the testing of underage children. If the test is for a treatable disorder, such as PKU or galactosemia, then it is clearly in the best interest of the child and the family to conduct the test routinely. Needless to say, this conclusion assumes that the disorder being tested for is not uncommonly rare and that the test is valid and reliable, with few or no false positive or negative results. The real problem will come with tests for untreatable disorders that show late onset, such as Huntington disease, or tests for the presence of mutations that have no direct effect on the child but that may affect his or her potential children, such as fragile-X syndrome. In these cases, there is no justifiable benefit to the child associated with early testing, and, in fact, the result of the test could adversely affect the manner in which the family regards, responds to, and raises the child. In these cases, it also makes no sense to obtain genetic information that would be withheld from the parents. Rather it appears prudent to wait until the child is of age and then let him or her decide if he or she wants to know this information.

Many will disagree with this conclusion, arguing that parents should and do have a right to know whether their child will develop a late-onset disorder or be at risk of having children with a serious disorder. Parents may wish to set aside a trust fund, for example, anticipating the needs of the child once the disorder develops in adulthood. Or the parents may want the information in order to make procreative decisions of their own. Whether we allow or prohibit selective testing of minor children, at the parents request, for untreatable genetic disorders that show late onset remains to be decided. This issue, though, clearly focuses our attention on the complexity of the ethical dimensions of genetic testing and on the importance of thinking, proactively, about the application of the new genetic discoveries.

9

Abortion and the New Genetics

Abortion is an unresolvable moral issue and so we do not try to resolve the issue in this chapter. But, by clarifying morality as it relates to abortion, we hope to assist practitioners in their abilities to counsel women and couples contemplating abortion of fetuses with genetic conditions. In particular, we show that it is important for a genetic counseling facility to develop a public policy concerning what prenatal genetic tests they will perform and when they will perform them. In order to aid in the developing of these policies, we offer a definition of a person as someone who has the full protection of morality and we develop a proposal for who counts as a person.

Introduction

Mr. and Mrs. W are returning to the Prenatal Diagnosis Center for genetic counseling. They have become friendly with the genetic counselor, as this is their third visit with her, and they have appreciated the support and guidance that she has provided. They did not plan this pregnancy and are experiencing both hope and dread, because their first two pregnancies ended with the birth of a son and then a daughter with a lethal genetic condition, Pena-Shokeir syndrome. This is a rare disorder in which the newborn cannot adequately breathe and dies within moments.

Mr. and Mrs. W's story began with the birth of their son, who, quite unexpectedly, had multiple joint contractures at birth and who did not have the lung capacity to sustain his life, even in the face of full Intensive Care Nursery efforts. It was for them an experience "from another world." They were counseled that their son had an autosomal recessive genetic syndrome, that each of them was a carrier of this gene, and that they had a 25% likelihood of recurrence with any subsequent pregnancy. They were told that this syndrome is thought to be the result of failure of normal brain and spinal cord development and function such that the fetus develops normally for some unknown period of time (at least until the second or third trimester), after which time the fetus can no longer move properly. Both brain and spinal cord function "fails." Consequently, the fetal limbs become tightly contracted and, more importantly, adequate lung development does not occur. Predictably, the lungs would not be capable of supporting the baby, and death at birth is inevitable.

Some time after the death of their son, Mr. and Mrs. W felt ready to enter into another pregnancy. Early in their second pregnancy, they sought counseling and were reminded that no genetic or biochemical test was available to indicate whether the fetus had Pena-Shokeir syndrome; amniocentesis and chorionic villus sampling were of no help. They were advised that fetal ultrasound examinations done throughout the pregnancy would, with a high degree of accuracy, identify a fetus that was not moving and consequently developing contractures. They were also told that the time at which the ultrasound identified an affected fetus might well be late in pregnancy, and perhaps after the time at which abortion could be offered by the practitioners of the Prenatal Diagnosis Center. In the course of their pregnancy, many ultrasound examinations were performed, each one reassuringly normal. However, at 28 weeks gestation, with Mr. and Mrs. W now themselves fairly skilled in ultrasound interpretation, it was apparent to them and confirmed by the ultrasonographers that this baby was affected with Pena-Shokeir syndrome. Her hands and feet were tightly contracted, excess amniotic fluid had accumulated, and the fetus lay quietly at the bottom of the uterine cavity. Now the couple was faced with the situation they had dreaded.

For Mrs. and Mrs. W, abortion was not an option. They expressed strongly held beliefs of opposing abortion. They had experienced the birth of their son and, although it was very difficult emotionally, they could not, in good conscience, actively end this pregnancy. They chose to continue the pregnancy. At the birth, they had gathered a few close family members to support them while their daughter died of complications of Pena-Shokeir syndrome.

As they return to the Prenatal Diagnosis Center, Mr. and Mrs. W ask the counselor a series of questions and are reminded that the technical information is unchanged—no molecular or biochemical tests have been developed; they must still rely on ultrasound examinations. In the course of this conversation, Mr. and Mrs. W wish aloud that a test did exist such that early prenatal diagnosis was conclusive. They tell the counselor that the emotional challenge of waiting so long to be reassured is overwhelming. They add that, if a test were able to tell them very early in the pregnancy that the fetus did have Pena-Shokeir syndrome, they would think seriously of abortion, with Mrs. W saying, "I don't know if I can bear to do it again" (carry on with a pregnancy and experience another newborn death).

This pregnancy was monitored in the same fashion as the second and, like the second, an affected fetus was identified at 26 weeks gestation. The pregnancy was continued and their second son died shortly after birth. Mr. and Mrs. W are now pursuing pregnancy by insemination of donor sperm.

This story is true and one that is faced by many women or couples and, at some time, by every prenatal genetics program. One can easily imagine another couple asking for, or even demanding, the abortion of a fetus with

Pena-Shokeir syndrome even though identification occurs late in pregnancy and beyond the traditionally accepted "viability" of the fetus. What is the duty of the genetic counselor in this situation? Is it morally acceptable to abort very late in pregnancy? How can the counselor assist in the abortion decision?

We are concerned with the topic of abortion in this book as it relates to the field of human genetics. The discussion of abortion for genetic conditions[1] will not, in all likelihood, moderate the passions surrounding the issues of abortion, but it may highlight some of the complexities of the concerns involved, particularly as it relates to abortion for reasons relating to characteristics of the fetus. It is our intention here to review the relevant moral issues of the topic of abortion and to propose a method (using an explicit account of morality) whereby practitioners such as genetic counselors or medical geneticists will be capable of clarifying their own position on abortion in specific clinical situations and, more importantly, will be more capable in assisting women or couples (the consultands) by making clear the implications of the choice of abortion. By clarifying morality as it relates to abortion, we hope to assist practitioners so that a woman or couple who is requesting an abortion for fetal genetic conditions is offered morally and legally acceptable options. Further, by bringing to light morally relevant features of decision making about abortion and by making explicit how morality relates to abortion, we intend to assist practitioners in their abilities to counsel women and couples contemplating abortion of fetuses with genetic conditions.

Abortion is such a controversial political and moral issue that it is difficult to discuss it without arousing such strong passions that the arguments are not attended to, and often there is little room for progress. On one side, there are those who hold that a fetus, from the time of fertilization, has a right to live that is sufficiently strong that only some very serious threat to the life or well-being of the pregnant woman justifies abortion. On the other are those who hold that the right of women to control their reproductive lives is paramount, and the choice to abort is theirs alone to make independent of any considerations of the status of the fetus.

Thus, most of the arguments that have been presented against abortion have been arguments that stress the benefit to the fetus of not being aborted, and most of the arguments for abortion have been those that stress the benefit to the pregnant woman of being free to have an abortion when she chooses and for whatever reason she chooses to have it.

In our discussion of this issue, we shall not discuss these arguments in any direct way. We hold that none of these arguments are conclusive—that is, that none of them are such that everyone must come down on one side of this issue rather than the other. We disagree with the extreme claims of both sides—those claiming that morality prohibits abortion in the same way that it prohibits the killing of adults, and those claiming that abortion is always and only a purely personal matter on which it is completely inappropriate to make laws that restrict the freedom of women to have abortions. Further, we realize that abortion is a topic about which reasonable moral people can, do, and will strongly disagree.

More important, there is no general agreement (or is it likely there ever will be) on the moral status of the fetus—that is, on the degree to which morality protects the fetus. Some hold that the fetus has the same moral status as any adult—that is, it is fully protected by morality; others hold that the moral status of the fetus is fundamentally different from that of an adult or a young child—that is, it either is not protected at all or is protected to a much smaller degree; still others hold that the moral status of the *early* fetus is different from that of the *older* fetus, whose status is more like that of a newborn.

In our discussion of abortion, we will use the explicit account of common morality provided in Chapter 2, but we will also enlarge on that account. In particular, we will discuss the scope of morality in far more detail than was done in Chapter 2. The reason for this is that questions about the scope of morality are central to the topic of abortion.

Whom Do the Moral Rules Protect?

Moral Agents

The question about the *scope* of morality—that is, who is protected by the moral rules prohibiting killing, causing pain and so forth—is one on which impartial rational persons can, within limits, disagree. Everyone admits that all those who are expected to obey the moral rules must also be protected by them. *Morality could not be a public system that applies to all rational persons if all rational persons were not protected by that system.* All competent adults and all children at or above the age at which they are expected to obey the moral rules (i.e., held morally responsible) are already included in this original group of rational persons who are moral agents. Briefly, moral agents are those who are held morally responsible for their actions. They must have at least the following characteristics: (1) some general knowledge germane to the moral rules[2] (for example, people can die, can experience pain and suffering, and so forth); and (2) some volitional ability, or ability to will (that is, they can act intentionally and are at least sometimes influenced in a rational way by knowledge of the consequences of their actions).

Are any beings in addition to moral agents protected by the moral rules? We must remember that infants and very young children are not moral agents; that is, they are not held morally responsible for their actions. Are they, justifiably, less protected? If so, why and what is their moral status? What about fetuses and nonhuman animals? Can having a severe genetic condition—that is, one that prevents the possibility of ever becoming a moral agent—affect the moral status of a fetus, or an infant, or even an adult?

Potential Moral Agents

Potential moral agents are those who will, unless something happens to prevent it, develop the characteristics of moral agents but who, for the moment, do not yet have those characteristics. All healthy infants and young children are potential moral agents, because they are likely to come to have those characteristics sufficient to be held responsible for their actions—that is, general knowledge germane to the moral rules and the relevant volitional abilities. In our society and in all developed societies, infants and children are fully protected by the moral rules; one cannot violate the moral rules with regard to infants and children without having the same justifications that one has for violating them with regard to adults—that is, actual moral agents.

However, if one holds that it is by virtue of being a potential moral agent that a human being has the status that deserves the equal protection of the moral rules that is afforded to all actual moral agents, then all fetuses *who are known not to lack the capacity to develop into moral agents* are of this status and deserve the equal protection of the moral rules. Those who hold that a fetus is protected by virtue of this potential then also hold that the fetus at any stage of development has the same moral status as that of a prematurely born infant, a two-month old, and a two-year-old—all of whom are protected by the moral rules and are potential moral agents. Because this status of potential moral agent is achieved from the time of fertilization, one who holds that potential moral agents are as such protected must hold that equal protection begins at fertilization. It follows, then, that such people must hold that forms of contraception that interfere with implantation of the fertilized ovum (e.g., intrauterine devices, Norplant, and some oral contraceptives) are morally unacceptable; that is, one is violating the moral rule "Do not kill" with regard to those who are protected by the moral rules.

There is significant disagreement about whether being a potential moral agent is sufficient to attain the status of being protected by the moral rules. It is obvious that reasonable people do disagree about whether one needs any justification for early abortion. Although some hold that a potential moral agent is a person in the same way that an actual moral agent is and, hence, hold that one needs as strong a justification for abortion, even in the first few weeks, as one would need for killing any person, others hold that one needs no justification at all for early abortion; still others hold that, even though in the first few weeks one needs some justification for an abortion, it can be much weaker than that needed for having an abortion when the fetus is older and is sentient.

Sentient Beings

Some hold that it is the status as a *sentient* being, rather than the status of potential moral agent, that gives one the equal protection of morality; that it is unjustified to limit the protection of morality to actual and potential moral agents. Sentience is the term used to indicate any capacity to experience pleasure or pain. Human beings who are capable of experiencing suffering are sentient beings. For example, the prematurely born infant, thought not a moral agent, is a sentient being, one who does experience pain and pleasure. A newborn with a lethal genetic condition or malformation, for example, who is not a *moral agent* and, perhaps, not even a *potential moral agent* in that he will not develop those capacities that define a moral agent, is nonetheless a sentient being who experiences pain.

If one holds that sentience itself is sufficient to give one full protection of the moral rules, then not merely older fetuses and newborn infants are protected by morality, but also many animals, especially mammals, are similarly protected by virtue of their sentience.[3] Although it is not at all clear when sentience develops in human beings—or even how to or whether one can ever empirically determine sentience in a fetus—all agree that the very early embryo is not sentient (i.e., cannot experience pleasure or pain), whereas the 39-week fetus is sentient in the same way a newborn is. Many hold that having the equal protection of the moral rules requires both sentience and being at least a potential moral agent. Thus, infants and fetuses in the later stages of development, being both sentient and potential moral agents, are protected beings; fetuses or embryos in the very early stages of pregnancy are not in that sentience has not yet been achieved. These people may hold that sentience in human fetuses then determines when such fetuses gain the protection of the moral rules without holding that sentience by itself is sufficient for having the protection of morality.

Those who hold that abortion should not be allowed except in the case of rape or incest are arguing that whether or not the fetus is sentient or of a moral status equal to that of an adult is either irrelevant or superseded by virtue of the circumstances of the pregnancy. They seem to be claiming that, if coercion or violence or both were involved or a woman is made pregnant by a family member, then abortion is allowed, but otherwise not. Those who argue in this way appear to be abandoning their arguments that the moral status of the fetus is the same as that of an adult, a status that affords equal protection from, among others things, killing without very strong justification. They cannot be arguing that the way the person came into existence determines whether or not it has the equal protection of the moral rules. For this would mean that children who were born because of rape or incest are not afforded the full protection of morality. It would seem to imply that even adults who have a similar origin do not have full protection.

If anyone holds both that the moral status of the fetus is equal to that of any adult, including the pregnant woman, and that abortion should be allowed only in the case of rape or incest, he or she must be holding that, in these two circumstances alone (rape and incest), the pain and suffering experienced by the woman in being forced to carry a fetus conceived by rape or incest justifiably override the status of the fetus as a protected being. This is a very implausible view. It undermines the claim that the status of the fetus is the same as that of any adult, because no one would hold that unintentionally causing that degree of psychological pain is sufficient justification for killing an adult human being. Holding both that the fetus has the same moral status as an adult and that *only* in the cases of rape and incest does the psychological pain and suffering of the pregnant woman justify abortion seems to involve an obvious inconsistency.

If pregnancy threatens the life of the woman, and abortion will save her life, most people hold that abortion be allowed, if not encouraged. Those who oppose all abortion except when the life of the mother is threatened seem to be arguing that the moral protection of the fetus is superseded by that accorded to the pregnant woman. This view seems to hold that there are two persons of equal moral status—the pregnant woman and the fetus—and because only one of them can live, one ought to save the one that has already established relationships with other moral agents, because saving that one will cause the least harm to other moral agents. In normal circumstances, it seems obvious that the mother would always be "chosen" in this circumstance. This argument is similar to that offered for picking certain persons to survive in "lifeboat" situations, and so it is generally felt that aborting the fetus to save the mother is morally acceptable and not a serious objection to the view that the fetus has the same moral status as an adult, while still allowing the life of the pregnant woman to justifiably override the life of the fetus.

Potential Sentient Beings

The most far reaching claim for equal protection by moral rules we can imagine is that which holds that all *potential* sentient beings are protected. Those who hold this view claim that the fetus, from the time of fertilization forward, is protected by morality on this basis. We remind the reader that this claim for protection by morality means *equal* protection. This position, similar to holding that all potential moral agents are protected, provides the protection of morality to all conceptuses and fetuses. But holding that all *potential* sentient beings are protected would also require holding that all mammal conceptuses and fetuses are protected including those of nonhuman primates and other mammals, because they are also potential sentient beings. It does not seem to us that many people actually hold this view.

As the group for which protection by the moral rules is claimed expands beyond that of moral agents to potential moral agents, sentient beings, and potential sentient beings, those who argue for the larger group will frequently modify their positions when they realize that protection by the moral rules means *equal* protection, meaning that the very early human fetus, mammals, and, for the largest group, even mammal fetuses have protections equal to that of adults (moral agents). Insofar as the arguments are based on premises that all people can understand and accept, it seems that neither of the extreme views—the group including all potential sentient beings on the one hand or only actual moral agents on the other—is a view actually held by very many.

Persons

If we mean by a person a being who has the full protection of the moral rules (i.e., the same protection enjoyed by a moral agent, or a healthy adult), then determining who counts as a person is not a merely verbal matter, but a point of moral significance. We are offering a proposal for who counts as a person that we think will be most acceptable to those actually involved in making, counseling, or carrying out abortion decisions on the basis of genetic testing. This proposal will not be one on which we expect complete agreement, because we know that, given that what we mean by the word "person," a person has the full protection of morality. Thus, it is an extremely serious matter to attempt to specify who counts as a person. But, in an effort to bring some thoughtfulness to the work of clinical geneticists, genetic counselors, and other medical providers, we offer the following discussion. We believe it may help those working with pregnant women and couples who are considering prenatal genetic testing of their fetuses and who are considering abortion for fetal genetic conditions.

Given that morality protects all persons, we suggest that "person" should include all and only sentient human beings who are, were, or (unless prevented) ever will be held morally responsible for at least one action.

We realize that there is some dispute about this specification of "person" and that some will restrict persons to those sentient human beings who are or were held morally responsible for at least one action—that is, *present or past moral agents,* but not *future or potential moral agents* (e.g., children). However, as mentioned before, we think that most people in the developed countries hold that infants and children are fully protected by morality; the only controversial addition is that the late fetus is also protected. There are also some controversial exclusions. Those in a persistent vegetative state do not count as persons, because they are no longer sentient beings. The same is true of infants with anencephaly. Even more controversial is the exclusion of infants with genetic conditions or malformations that are lethal early in life and for whom there is no successful treatment, because though they may be sentient, they will never be held morally responsible for even one action.

However, we do not hold that no protection should be afforded those who are not persons, only that the degree of protection of these beings is not necessarily equal to that provided to persons. Moreover, we do not even argue that *full* protection by the moral rules should be offered only to those who are persons. We realize that there are strong arguments that full, or nearly full, protection by the moral rules should be accorded to *any sentient human being,* even if it is known that the fetus or infant included in this category will not develop the ability to "ever be held morally responsible for at least one action"—for example, infants and children with genetic conditions that are lethal early in life and for which there is no successful treatment. Although we do not believe that equally strong arguments exist for providing full, or nearly full, protection to those human beings who are not sentient, we believe that there are strong arguments for prohibiting, for example, active killing of newborns with anencephaly or those in a persistent vegetative state without *very strong justification*—justification that is strong enough or very close to strong enough to justify killing a person.

However, we also realize that many hold views different from those expressed here and that they also have strong arguments for their positions. We think that, on our proposed specification of a person as a sentient being who was, is, or ever will be a moral agent, most would accept that all persons are entitled to the full protection of the morality. But we think that there will be considerable disagreement on how one should treat those who are not persons. Many, for example, argue that it is morally acceptable to abort a fetus with anencephaly at any time, because the evidence suggests that it is never sentient. Others hold that very late abortion is acceptable for fetuses with any conditions that are lethal in the newborn, even well after sentience is thought to be present. Further, there are those who hold that late abortions after sentience is present are morally acceptable when the fetus has any nonlethal, serious disability that makes it impossible for it to ever become a moral agent.

In any busy medical center, one can have the experience of knowing that in the Prenatal Diagnosis Center a fetus with a serious malformation discovered late in pregnancy is being aborted, while at the very same time a newborn infant (even one born prematurely) with the same serious malformation in the Intensive Care Nursery is afforded all intensive care efforts available, even occasionally against the wishes of the infant's parents. Can this be reconciled? We have diagnosed severe hydrocephalus in a fetus of 26 weeks gestation for which the woman subsequently received an abortion, while at the same time being consulted on a 32-week gestational aged newborn with the same severe malformation who was receiving full intensive care support. Our Genetics Center staff disagreed on the appropriate approach to these two situations. Some argued that severe hydrocephalus was similar to anencephaly—that sentience was not present due to the severity of the malformation—and that the fetus, in all likelihood, would never develop sentience let alone develop into a moral agent. They argued that, if the fetus was not aborted, newborn care was not likely to save the life of the infant; that is, this also was similar to anencephaly and so late abortion was morally acceptable. They claimed that, because the couple could obtain an abortion in another state (according to the policy at our Center, the gestation of the fetus exceeded the acceptable limit for abortion) and given the seriousness of this particular malformation, our Center ought to make an exception and perform the abortion. Others felt that the analogy to anencephaly simply did not hold and that we should assume sentience was present and that there was some, perhaps very small, chance that the fetus and the infant could develop those characteristics that define a moral agent. Some felt that the treatment of the newborn was "excessive"; others felt that the infant did not have a "lethal" malformation and deserved intensive care.

As the preceding example shows, it is very likely that allowing late abortion and nontreatment of newborns with anencephaly has "real world" consequences. The potential for extension to circumstances and conditions less well defined than anencephaly is very great. While we acknowledge that there is legitimate disagreement concerning whether and under what conditions late abortions should be allowed, we think that there is much less disagreement concerning the prohibiting of killing of newborns. We believe that it is necessary to explain why many are willing to allow late termination of pregnancy for anencephaly but not willing to allow killing of newborns with anencephaly. Aborting late fetuses is not, on a purely theoretical basis, different from killing newborns. Why then is it acceptable to abort a fetus with anencephaly but not actively kill (euthanize) a newborn with anencephaly? In the United States and many countries, it is not acceptable to actively kill newborns with anencephaly even when it is agreed there is no sentience, no successful treatment, and no potential to become a moral agent, yet many countries, including the United States, allow late abortions for anencephaly.

On the account of morality given in Chapter 2, violations of the moral rules are justifiable only if one can publicly allow such a violation. We believe that considering the consequences of everyone knowing that killing is allowed is different for fetuses and newborns. "In the real world," (1) the potential for extending the list of conditions for which one is allowed to abort late in pregnancy is far more limited than is the potential for extending the list of conditions for newborn euthanasia; (2) it is far less likely that there will be a social policy mandating that women have late abortions for certain genetic conditions than it is that newborns with that genetic condition be euthanized; (3) not allowing late abortion for serious genetic conditions creates a terrible burden for a particular person, the pregnant woman, whereas not allowing the killing of newborns for the same condition does not impose such a terrible burden on any particular person. We believe that these real world reasons explain why many would advocate publicly allowing late abortion of fetuses with serious genetic conditions and yet would not publicly allow killing of newborns with the same genetic conditions, and may not even allow nontreatment of these newborns.

Consider a couple with an infant with Tay-Sachs disease (TSD) who discover this diagnosis at birth (having been inadvertently uninformed about this possibility until the birth of the infant). TSD is characterized by normal growth and development until five months of age, at which time symptoms of visual loss, hyperacusis, and developmental delays occur. The infant goes on to lose the interest and ability to interact, to develop seizures and spasticity, and to die between ages two and four years. There is no successful treatment. The infant will die before becoming a moral agent. Very few would argue that the infant with TSD ought not be treated and cared for; even fewer would argue for or allow actively killing of the newborn with TSD. However, many would argue that late abortion for TSD is acceptable. That is, many would argue that, although theoretically there is no moral difference in aborting late in pregnancy for TSD and active killing of a newborn with TSD, the former is acceptable and the latter is not. The basis for holding this is that publicly allowing abortion protects the vital interests of another—the pregnant woman—and that the actual process of performing an abortion is much less likely to lead to unacceptable practices than the killing of the newborn. The latter, if found acceptable, could lead to a more widespread practice of killing newborns. Further, given that newborns with very serious genetic conditions may resemble healthy newborns very closely, active killing of newborns may cause significant mental anguish for those involved and may also contribute to diminishing our sense of belonging to a stable moral community that cares about all those included in it—namely, all persons.

We do not see any way of completely resolving the disagreement concerning the question of who should be accorded the same protection of morality as that accorded to persons. We think there will always be disagreement about the more general issue concerning who is protected by morality. We do, however, think that when it is made clear that we are concerned with that group that should be given *the same protection as moral agents,* there will be considerably less variation than there now seems to be.

We hold that human fetuses, without serious genetic conditions, after they are sentient should be considered "persons" and protected by the moral rules in the same way as a newborn infant. This does not separate older fetuses from infants, but it does separate the early fetus and embryo from the older fetus and infant. We feel that this view will be beneficial for clinical geneticists and genetic counselors to consider in their work. We think that scientifically determining that stage of development before which it is clear that the fetus is not sentient would have significant moral benefits, and not merely if one accepts our specification of persons. It would make clear that abortion at a stage earlier than this involves no pain for the fetus and thus would alleviate at least some of the discomfort experienced by those who are participating in such early abortions.

Everyone holds that all persons deserve the equal protection of the moral rules. Because all children who are sentient and potential moral agents are persons, all would naturally hold that such children are protected by the moral rules. The only group included in this specification of persons about whom there is any significant disagreement are fetuses in the later stages of pregnancy. Some few hold that the pregnant woman has the right to terminate a pregnancy at any stage, including that stage after which a fetus has become sentient, for any reason. More hold that the pregnant woman has the right to terminate a pregnancy at that stage after which a fetus has become sentient only for very strong reasons, but still for reasons that would not be adequate to kill all other persons. If we consider the couple discussed at the beginning of this chapter, we are describing an older fetus who is of an age such that sentience is likely to be present but who has a lethal genetic condition. We know that the fetus with Pena-Shokeir will not survive infancy and hence will never be a moral agent. Even though it is a sentient human being, this fetus is not a "potential moral agent" and hence on our account is not a person. Thus we would expect disagreement about whether or not it is protected by the moral rules. Although by our own specification this fetus is not a "person," it does not follow that it is afforded little or no protection of moral rules.

In dealing with fetuses that are not persons but have reached the stage at which the fetus is normally sentient, we think it extremely important that every genetic counseling center develop a public policy concerning abortion. First of all, we think it is crucial that every genetic counseling center make clear what specification of person it accepts or, rather, what fetuses it regards as having the full protection of morality. Then it must make clear what it will allow with regard to those late fetuses that are not persons: and here it may be useful to distinguish between those fetuses that are not sentient and those that are but will die in infancy. Although we think that most would allow abortion of fetuses with anencephaly at any stage, some would not allow this same latitude for sentient fetuses with lethal genetic conditions. However, many would argue that sentience is not enough to rule out abortion of the fetus in the third trimester. Many would reasonably argue that termination of a fetus with a known lethal genetic condition is acceptable *at any stage of pregnancy*.

Abortion and the New Genetics

The majority of abortions (99%) in the United States are done before the fetus reaches 20 weeks gestation; 91% are performed at 12 weeks or earlier. On average, women consistently report three reasons that lead them to choose abortion: (1) having a baby would interfere with work, school, or other responsibilities; (2) they cannot afford to have a child; or (3) they do not want to be a single parent or have a problem with their relationship with their husband or partner.[4] Many fewer abortions (approximately 1%) are done for reasons related to the fetus.

Prenatal diagnostic techniques[5] often can identify the fetus with structural malformations, chromosomal conditions, and many monogenic conditions that cause disabilities or chronic illness or both—what we call "fetal genetic conditions." It is the fetus with one of these conditions who is aborted following prenatal diagnosis for reasons that are not solely related to the life circumstances of the pregnant women—that is, the preceding list of reasons. Pregnancies for which prenatal diagnosis is utilized are typically desired pregnancies, but with a fetus who unexpectedly (or for known reasons is at a higher risk) is identified with one of a number of fetal conditions—malformations, disability, or chronic illness. These women and couples face somewhat different moral choices.

Two scientific endeavors will assist in resolving or lessening moral conflicts about abortion for reasons related to the fetus: an increase in the number of genetic conditions that can be identified by the use of molecular genetic techniques; and technological advances in fetal assessment such that early fetal diagnosis is available.

The earlier the diagnosis of genetic conditions the easier the moral choices for many women or couples. For example, consider the couple with the fetus with Pena-Shokeir syndrome. This is a genetic condition for which molecular diagnosis is not available because the gene for this condition has not yet been localized. If by virtue of new genetic discoveries this particular gene is mapped and cloned, early diagnosis (i.e., before sentience) will become available, using chorionic villus sampling or early amniocentesis and molecular genetic diagnostic techniques. Clinicians who have had to rely on late diagnosis by ultrasound could identify a fetus with this genetic condition as early as 12 weeks gestation—certainly before sentience and before the full protections of morality (in the view of many) are applicable. For this couple, the new genetic discoveries will have made a very meaningful contribution, for they are willing to consider early abortion but not later abortion, when the fetus has become sentient. They express the feeling that there *is* something different about the decision to abort early rather than later.

We believe that moral dilemmas for many will be prevented by the increased availability of early fetal diagnosis. For those who hold that the early fetus (i.e., "pre-sentience" or not yet a "person" by our earlier specification) is morally different from the older fetus, early identification of fetal genetic conditions will diminish the moral confusion of abortion. That is, if a women or couple hold the view that early abortion is a morally acceptable choice with little or no justification required, then for them the presence of a fetal genetic condition provides a completely sufficient justification. The presence of an unexpected genetic condition in the desired pregnancy is considered as strong a justification as, for example, the loss of a spouse or loss of income such that the pregnancy is suddenly seen in a different (less desired) way.

As the new genetic discoveries progress, it is likely that the number of genetic conditions that can be applied to early fetal diagnosis will increase. In addition, the number of conditions that are not typically thought of as severe or life threatening will increase. Given that many feel that no justification for early abortion is a morally acceptable view, one can imagine that couples will request prenatal diagnostic services from clinicians for genetic conditions of little or no effect.

If one holds that the early fetus is of a moral status different from that of the later fetus such that no justification for an early abortion is required, then it follows that one would also hold that women can choose such an abortion for reasons related to the fetus that may seem trivial to clinical geneticists, genetic counselors, and other medical professionals. Consider, for example, testing for gender alone followed by a request for abortion of all female fetuses. This is not an uncommon moral problem discussed by geneticists and genetic counselors. What is the source of moral discomfort? Why is there discomfort among the medical providers concerning abortion for gender, but not for Down syndrome or spina bifida? The answer may lie in the medical tradition of prenatal diagnosis. If, for the sake of argument, a women could identify the sex of her fetus in the privacy of her home without involvement of medical providers, she could ask for and receive an abortion early in pregnancy without any discussion with medical personnel about her reasoning. She would, thereby, avoid the entire negotiation and discussion with medical personnel who might hold that her reasons for abortion are unjustified. In this hypothetical situation the pregnant woman is making what is for her a morally acceptable decision to abort for gender—a decision that she has considered carefully and made as "comfortably" as is possible.

We believe that the question of the moral acceptability of abortion for gender arises, in part, from the fact that identification of genetic conditions in the fetus *requires interaction with medical personnel.* Gender is not thought of as a malady, whereas Down syndrome and spina bifida are, and it is providing genetic testing for nonmedical conditions that disturbs geneticists and genetic counselors. There seems to be much less of a moral problem with using genetic information and techniques to prevent maladies than with using them to create a more "desirable" child. It may be felt that, if one makes no distinction between using genetic testing to abort for gender and using it to abort for Tay-Sachs disease, then it may be more difficult to allow gene therapy for Tay-Sachs and not to allow it for gender or some other nonmalady condition such as eye color. (See Chapter 10 for further discussion of gene therapy.) Even though we do not think that new genetic discoveries will do away with the distinction between genetic maladies and genetic conditions that are not maladies, doing genetic testing for abortion for nonmaladies does make that distinction seem less significant. (See Chapter 7 for further discussion of genetic maladies.)

Although we hold that no justification is required for early abortion such that women who wish to abort on the basis of gender may do so, we also hold that it is morally acceptable for clinicians to refuse to provide prenatal diagnosis technologies to women for reasons the clinicians find unacceptable, when these are not standard medical practice. We believe that, as a result of the new genetic discoveries and advances in prenatal diagnostic technologies, this will be an increasingly complex problem for clinicians; that is, when a test becomes available for a fetal genetic condition, does it have to be offered? If the gene for Pena-Shokeir syndrome were mapped and cloned for clinical usage, it seems to us that all would offer prenatal diagnosis to our couple. But what if the genetic condition were, in the view of the clinicians, less serious or not even a malady? How does one decide what is sufficiently serious to warrant prenatal testing?

It is our view that morally relevant features of genetic conditions, as discussed in Chapter 8, may help to guide the clinicians. We would encourage clinicians to develop and publicly articulate guidelines about which conditions they will and will not test for and about situations in which testing will not be offered. There are two questions that are always asked by women when being counseled about a fetus with a genetic condition: (1) how does this affect the fetus (and me/my family); and (2) how effective is treatment, if any is available? These are morally relevant features of any genetic condition; the answers to these questions will determine, in many situations, what a woman will decide. It seems likely that there will be genetic conditions either that are of so little consequence (less serious) or for which treatment is so effective as to be curative that some clinical groups may decide not to offer testing for them. Many groups do not test for sex alone and for carrier status of genetic conditions.

For example, we were asked by a man with mild hemophilia A to provide prenatal diagnosis for this condition. With this X-linked genetic condition, no sons would inherit this gene from our patient; all daughters would. Our patient was a man with mild symptoms. His daughters, in all likelihood, would have no symptoms, because they would be carriers. He requested prenatal diagnosis specifically to abort all daughters because they would be carriers and he wanted to "stop this genetic disease in my family." We (and other centers) refused to perform the necessary tests on the basis that the daughters would be carriers with no symptoms, thereby making the testing unjustifiable. We believe that ours was a morally acceptable decision, based on the policies of not testing pregnancies for genetic conditions of no or very little consequence and of not testing pregnancies for carriers of genetic conditions when no consequences to the carrier are likely. Would it be morally acceptable to abort all female fetuses in this case? On the basis of moral status of the early fetus—yes; if no justification is required, then any justification is acceptable. Would it be morally acceptable to offer prenatal diagnostic testing in this situation? We believe it would be acceptable but, because the effect ("seriousness") of the genetic condition on the fetus is so little, it is the same as offering prenatal genetic testing *of the early fetus* for gender alone. It follows then that those centers that do offer testing for gender alone are behaving in a morally acceptable way. But just because it is morally acceptable to offer such testing, it is not morally unacceptable to not offer it. When reasonable people disagree, as they do on the need to justify early abortions, it is morally acceptable to either help people decide (prenatal diagnosis) or to refuse to offer such help.

Public Policy Statements

When, as in the question of which genetic tests should not be offered to a pregnant woman, the range of actions that are morally acceptable is quite broad, we believe it is important to make a public policy statement that describes the nature of testing offered by any prenatal diagnostic clinical group. In an effort to avoid moral conflict and at the same time inform one's prospective patients, we suggest that all clinical groups offering prenatal diagnostic services (1) develop the range of testing that they will offer, (2) develop the range of reasons for and timing of abortion services offered, and (3) publicize guidelines that articulate the principles that determine which tests will not be offered and when pregnancy termination will not be offered. As the new genetic discoveries progress, these principles can then be applied to new tests as they become clinically available. When necessary, the guidelines can be revisited and altered. Here is an example of one such policy statement:

> The XYZ Medical Center Prenatal Diagnosis Unit will offer testing for all genetic conditions of the early fetus appropriate to the patient's medical history, except those conditions for which there is, in all likelihood, no specific consequence for the fetus, or for those conditions in which treatment is curative. We do not test to determine if the fetus is a carrier of a genetic condition when there are little if any consequences for the fetus. On this same basis, we do not test for gender for the sole purpose of determining gender.

It is likely that any group of genetic and medical professionals will have a range of views on these topics. Likewise a group constituting a prenatal diagnosis "team" will have a range of views. Individuals within the group will probably disagree. It is unlikely that strongly held views will change. We suggest that groups engage in a process that expresses that range of views held by the team and that a discussion take place such that a policy statement is produced by each group. The process of examining the views held by team members may serve to benefit the working relation, and, importantly, will end with a policy statement that has had open discussion. We are not suggesting that consensus be achieved; but we do feel that the group determine the borders of what they hold to be morally acceptable behavior with regard to prenatal genetic testing and abortion for genetic conditions. We would then expect to see guidelines and policy statements that differ somewhat from one center to the next, but we would also expect to see more similarities than differences and would expect that all would be morally acceptable approaches. We think that providing the clients with such public policy statements is an important part of the informed-consent process.

Earlier in this chapter, we offered a specification of "person" that has significant moral implications, because a person has the full protection of morality. Because we include as persons all those sentient beings who are, were, or ever will be moral agents, what will be the effect of the new genetic discoveries on abortion of the fetus after sentience has been established? If the fetus does not have a genetic condition that will prevent it from becoming a moral agent, then it is a person, and abortion after personhood has been established is morally prohibited without very strong justification—for example, the life of the pregnant woman is threatened. However, if genetic testing shows that the fetus or infant will not develop the ability to "ever be held morally responsible for at least one action" and thus is not a person according to our specification, then the issue is a matter to be decided by each genetic counseling center. As mentioned before, there are strong arguments that equal, or nearly equal, protection by the moral rules should be accorded to *any sentient human being, including a fetus,* even if it is known that it will die in infancy. But there are also strong arguments that the interests of actual persons completely outweigh the interests of nonpersons, especially those that are not even sentient, but also those that are. We do not see any benefit to be gained from putting forward any controversial position; our goal is to lay out clearly and explicitly what we believe to be the limits of the morally acceptable positions.

The new genetic discoveries have the potential to lessen the need for consideration of late abortion by virtue of the development of early molecular diagnostic techniques that can identify such conditions before sentience and so before personhood. This will avoid for many the moral conflicts of pregnancy termination. We know that the issue of the moral acceptability of late-pregnancy termination will still remain for many conditions for which the new genetic discoveries offer little promise in changing from a late (mid-trimester) diagnosis to a very early one—for example, renal agenesis or anencephaly. Both of these conditions are diagnosed late in pregnancy using ultrasound technology; both are examples of conditions that are lethal in the newborn period and that are regularly diagnosed after fetal "viability" and sentience, if it will ever arise, will have been established. As such, they represent conditions being discussed by some authors as justifications for late-pregnancy terminations. We feel that the new genetic discoveries can be of great benefit to women and clinicians in preventing late diagnoses, thereby preventing troubling moral conflicts relating to abortion. But with this apparent progress, existing moral conflicts may be exaggerated—that is, our becoming a society that demands the use of genetic testing of the fetus for conditions associated with any illness or disability or both.

In this chapter, we have made the following points: (1) that the moral status of the early fetus is different from that of the older fetus (after sentience) such that it is morally acceptable to hold that no justification is required for early abortion; (2) that most late abortion is morally prohibited without very strong justification; (3) that the new genetic discoveries will increase the ability to test for genetic conditions including those in which the effect of the genetic condition is inconsequential or easily treatable; (4) that the new genetic discoveries and advancement in technology will increase the ability of early fetal genetic diagnosis; (5) that the range of morally acceptable testing of the early fetus has few restrictions, but that it is also morally acceptable not to test for nonmaladies; and (6) that the public interest would be well served by prenatal diagnosis groups developing public guidelines regarding prenatal genetic testing and abortion for fetal genetic conditions.

This suggests that, as the new genetic discoveries progress, there may well be more women requesting abortion for reasons related to fetal genetic conditions and, plausibly, that there will be more conditions identified by the new genetic discoveries that are not serious but are considered "mild" or inconsequential for which women ask testing. The larger social question then becomes, in part, one of the effect of such testing on women and on all of society. Further, the effect of the new genetic discoveries on individuals with genetic conditions and their families needs careful consideration. Will women or couples who give birth to an infant with a genetic condition be looked upon negatively because they did not use (or were not aware of) available genetic testing? Will family planning result in some couples being isolated or discriminated against by their choice not to engage in testing? Are we going to see a "new eugenics,"one in which the information gained from the new genetic discoveries and the advancements in early fetal diagnosis lead to women gaining greater control over their reproduction with the consequence that they have fewer births of children with genetic conditions, but leaving those not aborted to experience greater discrimination? How can we even consider limiting the choice of genetic testing of the early fetus without limiting the rights of women, particularly given the foregoing discussion concerning the moral status of the fetus?

One might argue that there can be no "down side" to women having the opportunity to know more about the possibility of genetic conditions affecting their fetuses. But as pointed out in Chapter 7, we know that more knowledge is not always beneficial to the person who has it. On the surface, it appears that women and couples will have more choices and, hence, more control over the genetic constitution of their offspring. But, given the complexity of these choices, for such knowledge and control to be used appropriately by women and couples requires new information and new supports for women and couples making such choices.[6] The informed geneticist and genetic counselor can be important sources of information and support. In order to provide both, we feel it is important for such professionals to have a sense of their own views on the issues discussed in this chapter, because then they can provide appropriate support and guidance for women and couples who are their clients.

Endnotes

1. "Condition" is used because—along with disease, disorder, and defect—it is common language for practitioners; it has the same meaning as "genetic malady" which is used elsewhere in this book but is not typical language for practitioners.

2. See Chapter 2 for a discussion of moral rules.

3. This might be the view, for example, of those who hold that animals have the same rights as human beings.

4. *Facts in Brief: Abortion in the United States* (1990). The Alan Guttmacher Institute, New York.

5. For this chapter, we mean all techniques currently available, including chorionic villus sampling, amniocentesis, ultrasound, biochemical and molecular genetic testing.

6. For a discussion of the complexities of decision making by pregnant women utilizing prenatal genetic testing see R. Gregg (1993): *Pregnancy in a High-Tech Age: Paradoxes of Choice.* Paragon House, New York.

10

Ethics of Gene Therapy

In an ideal world, where all scientists can be depended on to observe the strictest standards before engaging in germ-line gene therapy, all of the arguments against developing and using it have serious flaws. It does seem that one can draw a non-arbitrary line between negative and positive eugenics, and there is not even any good theoretical argument against positive eugenics. Germ-line gene therapy will not deplete the gene pool or adversely affect evolutionary development. However, in the real world, some scientists will exaggerate the benefits and minimize the risks. Germ-line gene therapy produces few benefits that cannot be attained in a less risky way. In the real world, these benefits do not warrant taking even small risks that could have long-term disastrous consequences.

Introduction

When the British Royal Navy instituted the practice of providing all sailors at sea with a daily ration of limes as a measure to prevent scurvy, it became the first government agency to deploy "replacement" therapy for a species-wide genetic deficiency. Human beings, unlike goats or cows or horses or most other mammals, are metabolically unable to synthesize directly their own minimal daily requirement of vitamin C and will, most likely forever, require it as a supplement to their diet. But vitamin C deficiency is not a genetic malady, because, as discussed in Chapter 7, in defining malady the notion of distinct sustaining cause is invariably defined relative to the norm of the species, which in this case is a universal lack of production. Replacement therapies for real genetic maladies are currently used in a number of cases. In most situations, affected individuals receive a dose of the biologically active gene product, such as insulin for insulin-dependent diabetes or clotting factor protein for classical hemophilias. Alternatively, affected individuals may receive a dose of a small molecule, such as vitamin D or vitamin B_{12}, when a genetic mutation has eliminated the organism's ability either to utilize or to synthesize that molecule directly.

Unfortunately, the vast majority of genetic maladies cannot be treated by conventional replacement therapy—that is, by providing a dose of the normal gene product. Usually the problem is that the normal gene product or metabolite is required within a small number of specialized cell types that are often inaccessible, such as liver or brain cells, or it is required for a brief but very specific period of time. Often, this time is when the organism is an embryo or fetus and is developing in the inaccessible reaches of the maternal womb. In these cases, even the isolation of a purified, normally functioning gene product or small molecule is of no use therapeutically, because there are no means of delivering those proteins or metabolites to the right place at the right time. The technology is changing, however, so that it will be possible to use conventional replacement therapy to treat embryos or fetuses for a variety of genetic maladies, but right now such technologies are limited to fetal surgery, tissue transplantation, and blood transfusion. In the 1970s, however, a completely novel approach to replacement therapy began to develop as the science of molecular biology gave birth to recombinant DNA technology.

During the 1970s, geneticists began to devise methods for isolating single, identifiable genes from human beings and for inserting those genes into a foreign chromosome. Initially, gene cloning, as it is termed, was restricted to inserting specific genes from a plant or animal source into a bacterial plasmid or into a viral chromosome. Once inserted, these genes would replicate along with the rest of the organism's genome and often would be expressed biologically, so that the inserted gene became transcribed into a functional messenger RNA that is then translated into a functional protein. The first important protein produced in this way was human insulin, but now there are dozens of such recombinant proteins available including human and bovine growth hormone and tissue plasminogen activator, a protein useful in the treatment of heart attacks. As the technology reached higher and higher levels of sophistication, new opportunities for the application of this recombinant DNA technology arose in the areas of pharmaceuticals, energy, and agriculture. In agriculture, for example, recombinant DNA technology has produced crop plants resistant to herbicides and insect attack, as well as tomatoes with a longer shelf life. Based on the self-evident or speculated commercial application of these techniques for cloning, sequencing, and expressing genes, a new commercial industry, designated genetic biotechnology, was created.

It rapidly became clear that the same genetic biotechnology that was being employed to modify the heredity and biological function of bacteria, yeast, fruit flies, tomatoes, tobacco, and mice could also be developed for and applied to human beings. After the discovery and the physical isolation of several human genes involved in serious maladies, many geneticists came to believe that inevitably genetic transformation would be tried on people at first to correct severe hereditary defects such as thalassemia, severe combined immune deficiency, or cystic fibrosis. A major ethical concern was that either somatic or germ-line gene therapy would not only be applied toward worthwhile goals in medicine, but also be developed for the nonmedical use of human genetic engineering. Many came to believe that the introduction of gene therapy for the purposes of negative eugenics—that is, to treat or eliminate serious genetic disorders—would pave the way for the introduction of positive eugenics—an approach directed toward enhancing, improving, or perfecting human beings. This might include the inappropriate use of the technology to produce enhanced size, strength, aggressiveness, or intelligence.

Although the concern about the use of gene therapy for nonmedical reasons was obviously a form of the classic "slippery slope" argument, no convincing argument has been provided that shows that somatic cell gene therapy involving human beings is, in any way, socially or ethically unacceptable. Many discussions have concentrated on the use of somatic cell gene therapy, even though the added gene is not inherited by offspring of the patient being treated. But no argument has been provided to show why somatic cell gene therapy differs in any morally significant way from cosmetic plastic surgery. Although one may be concerned about limited medical resources and so prefer that plastic surgery be used to repair serious problems, no one claims that it is unethical to use plastic surgery to further enhance the appearance of those without serious problems. A far more important problem is raised by the development and use of germ-line gene therapy in which eggs, sperm, or very young embryos are transformed genetically and the gene is inherited by offspring of the patient being treated

In this chapter, several of the ethical issues surrounding positive and negative eugenics, involving somatic cell and germ-line gene therapy, are discussed and evaluated. We will endorse the use of somatic cell gene therapy, subject only to the same constraints that govern the introduction of any powerful therapeutic procedure that poses significant risk. Although we will, in the end, discover no theoretical reason for not using germ-line gene therapy, we will provide persuasive arguments, based on real world considerations, that lead us to propose a moratorium on the development of any form of germ-line gene therapy involving human beings.

State of the Art

We have discussed in early chapters the relation between genes, proteins, phenotypes, and genetic maladies, and so a detailed discussion of heredity is unnecessary. Although many genes are polymorphic in human beings, containing from two to thirty or more different alleles in the species, in most cases these heritable polymorphisms are ubiquitous and lead to phenotypic differences that are of little or no consequence for an individual. The ABO blood type series is a good example of a polymorphism for which there is no known important difference in biological fitness between those individuals who are of A, B, AB, or O blood type. However, certain allelic variants, usually rare ones, that are found at frequencies of less than one in a thousand do produce deleterious effects on the phenotype and are consequently thought to produce serious genetic maladies in the medical sense. In the case of hemophilia A, the affected individual has inherited only the defective form of a gene whose normal function is to produce a blood protein crucial for clotting.

Some of the genetic polymorphisms, such as those involved in height or skin color in human beings, are not inherited as discrete alternatives, but rather show a continuous range of phenotypes in the species. Such continuously varying phenotypes or traits, whose expression is based on the interaction of many different genes, will demonstrate a complex pattern of inheritance. Moreover, these phenotypes or traits are often affected greatly by variations in the environment. Although there are likely to be many medically important genetic disorders that are shown to be multifactorial (polygenic and influenced by environment), such traits that show a complicated genetic basis will not likely be considered candidates for gene therapy soon: rather, deleterious traits showing a simple recessive pattern of inheritance, involving a single gene, will be likely candidates initially.

The development of the ideas and procedures that form the basis of human gene therapy can be traced to the early 1940s when Avery, McLeod, and McCarty studied the nature of DNA-mediated gene transfer, then called transformation, in the microbe pneumococcus. The first attempts to transform, genetically, somatic mammalian cells with free DNA took place in the 1960s and involved rodent and human cell lines that were genetically deficient in their ability to produce the purine salvage pathway enzyme hypoxanthine guanine phosphoribosyl transferase, or HGPRT. Although purines are known to be involved in the structure of DNA, in human beings HGPRT enzyme deficiency constitutes a well-known but extremely rare genetic malady called Lesch-Nyhan syndrome. Patients afflicted with this malady suffer severe mental retardation, a plethora of physical abnormalities, and, most characteristically, uncontrolled self-mutilating behavior. The malady is invariably lethal. However, the early in vitro attempts to transfer the normal HGPRT gene into these mutant cells were largely unsuccessful, because the scientists lacked the sophisticated technology that is now available. Today geneticists can easily insert the normal form of any human gene into a vector (plasmid or virus) and then use it effectively and efficiently to deliver the normal gene into the cell in a manner that leads to its stable integration into a chromosome and, ultimately, to the gene's timely and cell-specific expression. The one attempt, in the 1970s, to cure a human genetic disorder, hyperargininemia, involved injecting shope papilloma virus into two severely disabled patients. The notion was that the viral gene encoding the enzyme arginase was equivalent to the missing human form of arginase and could genetically "rescue" the patients by becoming inserted into a chromosome in the target cells, thought to be hepatocytes. The experiment failed, with little new knowledge acquired, and no cure produced.

During the 1970s and 1980s, the reagents and techniques required for genetic biotechnology were assembled. A large number of normal and functional human genes were cloned and sequenced and were later expressed in mammalian cells, using DNA or viral-mediated gene transfer procedures. By the mid-1980s, the possibility of carrying out effective somatic cell gene therapy on human beings became a realistic goal. Certainly, this sense was reinforced by the enormous success that other geneticists were having in carrying out somatic cell and even germ-line transformation, using a number of invertebrate and vertebrate species. The precedent setting "Mighty Mouse" experiment, in which the human gene for growth hormone was used to genetically transform a mouse embryo (a form of germ-line gene therapy) created an adult mouse twice normal size. What was most significant was that this was a germ-line procedure and that therefore the trait for very large size was passed on to Mighty Mouse's offspring.

The Issues Raised

It is generally agreed that somatic cell gene therapy for serious genetic maladies, involving the transfer of functional genes into somatic cells, poses no new ethical problem. This is, after all, simply an extension of routinely used replacement therapies (discussed earlier). There appears to be no important moral distinction between injecting insulin into a diabetic's leg and injecting the insulin gene into a diabetic's cells. In a practical sense, however, a distinction is clear: in principle, the insulin gene need only be injected once for lifelong therapy. Although the first application of somatic cell gene therapy in human beings, carried out by Martin Cline in 1980, created a great deal of controversy, more recent studies on nonhuman animal models and several NIH approved clinical trials using human patients have demonstrated both the feasibility and the importance of this approach. The introduction of recombinant retroviruses as efficient vectors for infecting target cells and introducing foreign DNA increased the range and specificity of somatic cell gene

Table 10.1 Gene Therapy Trials

Disease	Gene Inserted	Principal Investigator
AIDS	Thymidine kinase	Phil Greenberg, Univ. of Washington
AIDS	HIV env	Douglas Jolly, Viagene
ADA deficiency	Adenosine deaminase	Michael Blaese, NCI
ADA deficiency	Adenosine deaminase	Dinko Valerio, Univ. Hospital, Leiden
ADA deficiency	Adenosine deaminase	Claudio Bordignon, Lab. of Hematology, Milan
Advanced cancers	Tumor necrosis factor	Steven Rosenberg, NCI
Advanced cancers	Tumor necrosis factor	Steven Rosenberg, NCI
Advanced cancers	Interleukin-2	Steven Rosenberg, NCI
Brain tumor	Thymidine kinase	Kenneth Culver, NCI
Cancer	Interleukin-4	Michael Lotze, Univ. of Pittsburgh
Hemophilia B	Factor VIII	Jerry Hsueh, Fundan Univ., Shanghai
Kidney cancer	Interleukin-2	Eli Gilboa, Sloan-Kettering
Liver disease	HDL receptor	James Wilson, Univ. of Michigan
Lung cancer	Antisense ras/p53	Jack Roth, M.D. Anderson
Malignant melanoma	HLA-B7	Gary Nabel, Univ. of Michigan
Malignant melanoma	Interleukin-2	Eli Gilboa, Sloan-Kettering
Neuroblastoma	Interleukin-2	Malcolm Breener, St. Jude's, Memphis
Ovarian cancer	Thymidine kinase	Scott Freeman, Univ. of Rochester

therapy, and retrovirally transfected transgenes have now been shown to successfully correct genetic defects in a large number of cell types, including bone marrow stem cells, fibroblasts, myoblasts, hepatocytes, vascular endothelial cells, and respiratory airway epithelial cells. The genetic and nongenetic defects involved are summarized in Table 10.1.

Gene therapy, involving gametes, fertilized eggs, or early embryos, leads to germ-line transformation. Not only is the therapy permanent and continuing during the entire lifetime of the affected individual, but the transgene also becomes heritably transmitted to future generations, creating normal individuals who might otherwise have never existed because one of their parents would not have survived to reproduce. The major concerns that have been raised against germ-line gene therapy fit into three categories. First is the "slippery slope" argument, which is that we should not proceed because there is no unambiguous distinction between negative eugenics, the elimination of deleterious genes, and positive eugenics, the systematic improvement or perfection of the human genome. The unexamined assumption of this argument is that positive eugenics is unethical and undesirable. The second major concern is that by eliminating some deleterious genes we will lose some important genetic variation that may have a future survival value to the species. Sickle cell anemia is an example frequently cited because where malaria is prevalent the sickle cell mutation confers a survival advantage on heterozygous carriers. The third concern is the iatrogenic one that, by carrying out germ-line gene therapy to cure serious genetic maladies for present persons, scientists may cause even more serious maladies in future generations. In the remainder of this chapter, when we talk of gene therapy we mean germ-line gene therapy, unless explicitly stated otherwise.

Positive and Negative Gene Therapy

Following eugenics terminology we can distinguish two types of gene therapy. Procedures used to cure or treat serious genetic maladies, such as sickle cell anemia, are termed *negative gene therapy,* whereas procedures intended to improve or enhance traits, such as intelligence or endurance, in an otherwise normal individual are designated *positive gene therapy.* The argument put forth by some is that, if we use the technology to carry out negative gene therapy to cure a serious genetic malady, we will be unable to draw a nonarbitrary line that prevents positive gene therapy. The implication is that, because we are unable to draw a nonarbitrary line between negative and positive gene therapy, we should protect ourselves against the latter by not even beginning with the former. Our position is that one can draw such a nonarbitrary line, which distinguishes positive from negative gene therapy, by defining what a genetic malady is. Genetic conditions, such as hemophilia, cystic fibrosis, and muscular dystrophy, all share features common to other serious maladies, such as cancer, and fit the definitional criteria of malady (see Chapter 7). Thus, an objective and culture-free distinction can be made between genetic conditions that count as maladies and those that do not. A genetic condition that does not meet the definitional criteria of a malady should obviously not be counted as a malady, and gene therapy for such conditions constitutes positive gene therapy. Examples of nonmaladies might include blue eyes, widow's peak, freckles, O blood type, and curly hair.

As stated in Chapter 7, "A person has a malady if and only if he has a condition, other than his rational beliefs and desires, such that he is suffering or at increased risk of suffering a harm or an evil (death, pain, disability, loss of freedom or opportunity, loss of pleasure) in the absence of a distinct sustaining cause." This definition of malady, although it recognizes the normative character of a malady, is based on universal and objective factors, not on cultural and subjective ones. Further, it is not technical and so should be easily understood by everyone. Because it picks out almost all of the same conditions as all of the other widely used definitions but is more precise and has no clear counterexamples, it should have wide appeal. Nonetheless, as indicated in Chapter 7, it is inevitable that the definition will be vague to some small extent and that there will be some genetic conditions about which there will be disagreement concerning their malady status. We believe that the number of such conditions will remain small and that the disagreement is due to the nature of maladies, not to an avoidable vagueness in the malady definition. Instead, borderline conditions will be conditions about which people disagree on their malady status, such as short stature or mild obesity, because it is not clear whether the harms suffered owing to these conditions are universal or are primarily due to cultural conditions. Because such borderline conditions are not very serious in the medical sense, they are quite unlikely to be candidates for gene therapy, at least initially. For all practical purposes, there is no problem in distinguishing between negative and positive gene therapy, because there are few borderline cases and even fewer that are likely candidates for gene therapy.

However, whether a nonarbitrary distinction can be drawn between positive and negative gene therapy is not the fundamental issue; the concern rather is whether positive gene therapy is in some way an unethical practice. If it is not, then our society may decide that gene therapy used to enhance a normal child's intelligence, performance, stamina, resistance to toxic wastes in the environment, or even appearance is not unacceptable. It is important to remember that we are now talking about positive germ-line gene therapy. Unless some argument is provided to show that somatic cell gene therapy has serious risks, it does not seem that there is any stronger reason not to have positive somatic cell gene therapy than not to have plastic surgery to improve the appearance of normal people. Indeed, it is hard even to imagine what such an argument against somatic cell gene therapy could be. The case is somewhat different with germ-line gene therapy. We could conclude that germ-line gene therapy for even serious genetic maladies creates unacceptable risks to future generations. However, in the absence of convincing evidence that positive germ-line gene therapy poses such unacceptable risks, some claim that there is no more argument against positive germ-line gene therapy than against positive somatic cell gene therapy, cosmetic plastic surgery, or the taking of vitamins to enhance one's physical condition.

Evolutionary Consideration

Several critics have argued that those deleterious alleles that will become systematically eliminated by gene therapy procedures may be of some future benefit to the species. They argue, by analogy, that the trend in agricultural crop management this century to develop highly inbred, genetically uniform and highly productive strains of crop plants has created problems. In two cases, the macaroni wheat rust infestation of 1954 and the corn blight fungus of 1969, a new pathogen nearly decimated the inbred American crop and new genetic strains, resistant to the pathogens, had to be selected from outbred seed collections maintained in a seed bank. The critics argue that, had there been no effort to maintain the naturally occurring genetic variation lost by selection, the two species might have gone extinct. The genetic variation of a species, then, affords evolutionary plasticity or potential for subsequent adaptation to new and perhaps unforeseen conditions. To eliminate a deleterious mutant allele, such as that resulting in cystic fibrosis or sickle cell anemia, has or could have some risk. Recall that sickle cell anemia almost certainly evolved as an adaptive response to malaria.

The argument is false for two different reasons. First, one must consider the nature of genetic maladies. For maladies based on the inheritance of recessive alleles, it is not the presence of two mutant alleles that causes the malady, it is rather the absence of a normal allele (see Appendix, Figure 6). As long as a normal allele is present, the mutant alleles produce no effect. The situation of heterozygote advantage, or heterosis, as in sickle cell anemia, is not very common. So, in principle, gene therapy for recessive disorders will work as long as one normal allele is introduced. The mutant and nonfunctional alleles may still remain. This brings up the major point. Although selective inbreeding of crops leads to favorable strains at the price of genetic diversity, gene therapy, as it is currently done, does not lead to a loss of allelic variation. In fact just the opposite occurs. Functional alleles are added to a gene pool that contains the mutant form.

Currently, there are no procedures for human beings that allow the gene therapist to replace the defective allele with a functional form by homologous recombination. Such a capability has been developed for mice, by Cappecchi at Utah and by Smithies at Wisconsin: analogous procedures are being developed for rhesus monkeys and might soon be available for human beings. So, although it is not yet possible to *replace* a nonfunctional mutant allele, there may well be such a procedure available soon. Gene replacement procedures will expand the range of candidate genetic maladies subject to gene therapy by including maladies caused by dominant alleles.

Realistically, it seems quite unlikely that there will be any serious attempt to eradicate a deleterious allele from the human gene pool, even if it becomes possible and desirable. The technology required will be very expensive, at least initially, and as a consequence will be applied on an individual basis, with rather limited accessibility. Because it is a surgical procedure, germ-line gene therapy would be done in a medical setting and on a voluntary basis. So, although many couples might qualify for gene therapy, only a small number would elect to participate. Finally, it is well known that the vast majority of deleterious alleles that are recessive are maintained in heterozygous condition by carriers. Because there would be no reason to perform gene therapy on heterozygotes, the frequency of deleterious alleles would still be maintained at high levels. For example, if germ-line gene therapy involving gene *replacement* could be developed for Tay-Sachs disease and was used to treat all homozygous Tay-Sachs embryos (which occur at a frequency of 1/2000), the frequency of the Tay-Sachs allele in the entire population would decrease only from 0.01000 to 0.0099 in one generation.

Iatrogenic Risks

Germ-line gene therapy would be carried out on embryos in order to prevent a future person from having a serious genetic disorder, rather than to cure the disorder in the person after it has manifested itself. The therapy not only benefits an organism who would otherwise be an affected person, it also benefits that person's potential offspring. Because embryos and their potential offspring cannot provide or deny consent for the gene therapy, the couple contemplating gene therapy serves that role. If there were no risks at all associated with germ-line gene therapy, the issue of informed consent would pose no problem. If the risks were limited to the person the embryo may become, then there would be no more problem than that involved in parents choosing surgery for their infant children. However, it is known that gene therapy carries with it several known risks and that these risks may affect all of the offspring of the person for all generations.

Although there are several unknown risks, the major known risk emerges from the fact that the transgene is not targeted with respect to the chromosomal location into which it becomes integrated. Therefore transgenes become capable of causing what is known as insertional mutagenesis. Insertional mutagenesis, associated with DNA- or viral-mediated gene transfer, has been documented on a number of occasions. Presumably, the insertion of foreign DNA may disrupt some native gene function should the site of insertion be within or near the affected gene. If the insertion leads to a recessive mutation, no problem would arise unless that mutation were somehow made homozygous. Insertions could also produce mutations that act in a dominant fashion, in which case the fetus, if it survived, would express the malady associated with the insertional mutation event during his or her lifetime. Among the more serious consequences of disrupting the normal function(s) of a native gene by transgene insertion is cancer. There are a number of well-described human genes, collectively called proto-oncogenes and tumor suppressor genes, that perform normal and necessary functions that include the regulation of cell division. An obvious doomsday scenario has the transgene insert itself within or near a proto-oncogene or tumor suppressor gene and as a consequence the proto-oncogene or tumor suppressor gene is expressed abnormally, in time or space, or not at all. This points to the urgent need to design protocols leading to a more precise targeting of transgene insertion, to "safe" sites in the chromosome. Equally important will be protocols that ensure the stable and permanent integration of the transgene, so that the transgene does not become relocated later in the life of the individual or in subsequent generations. The major issue raised is that, even if germ-line gene therapy becomes an acceptable goal, there must be precise guidelines that reflect the limitations of the technology—limitations that could undeniably restrict the use of germ-line gene therapy to treating only very serious, life-threatening genetic maladies. Such strict standards provide further responses to the "slippery slope" argument.

Real-World Considerations

Based on the analysis of risks and benefits, germ-line gene therapy would be justified only in cases of severe maladies, where there is no less radical way of achieving the same goal. Among the class of less radical procedures would be somatic cell gene therapy or any other nongenetic therapy that would achieve the goal of preventing or curing the malady. For instance, one could provide the missing or defective gene product to the cells in which it is required and at the appropriate time. A more recently developed alternative involves preimplantation genetic screening, in which embryos are first produced by *in vitro* fertilization. At an early blastocyst stage of development, when the embryo is at the 16- or 32-cell stage, a single embryonic cell is removed and screened genetically for the presence of normal or defective alleles, using the powerful technology of polymerase chain reaction (PCR). If PCR analysis reveals that the embryo will not develop into a fetus that is affected by the genetic malady, uterine implantation would be carried out and normal development would take place. If PCR analysis revealed that the fetus would develop the genetic malady in question, the embryo would be discarded before implantation.

Discarding an early embryo before implantation does not pose a serious problem for most people without the relevant religious or metaphysical beliefs. (See Chapter 9 for a more detailed discussion of abortion.)

Consequently, preimplantation screening techniques, which can identify malady producing genes in a 16 cell embryo, eliminate the need for germ-line gene therapy entirely. Germ-line gene therapy, then, will only be useful for improving or enhancing people by adding new genes for strength, vigor, performance, or for resistance to pathogens or toxins. There will be cases in which both parents are homozygous for a rare deleterious recessive allele, such as cystic fibrosis, but the number of such cases will be vanishingly small. Because the primary use of germ-line gene therapy will be to add improvements rather than to eliminate defects, it may give rise to serious social and political problems. As pointed out earlier in this chapter, gene therapy will be, for the foreseeable future, a very expensive procedure. Thus, only the wealthy will be able to afford it. This makes it more likely that the economic and class differences we now have will become hereditary differences.

Although wealthy families can now pass down their money, thus perpetuating social and political power, this method of perpetuating family power and influence does not always work. The next generation may not be capable of maintaining the family fortune. Even providing access to the best schooling and other advantages does not guarantee success. Germ-line gene therapy probably comes as close as is humanly possible to guaranteeing that those families who can afford it will be able to perpetuate their social and political dominance. Thus it may give rise to a genetically stratified society as envisioned in *Brave New World.* Even more speculatively, once this technology is well developed, it can be used by those societies in which those in power are not governed by ethical restraints. People may be genetically engineered to perform various tasks—for example, as warriors. Imagine a group of people engineered to be resistant to various poisonous gases—for example, sarin. It does not take much imagination to see what problems might arise.

However, these concerns, although genuine, are speculative. On the other hand, we know from experience that cutting edge technology, including genetic technology, generates pressures for its use. Consequently, it is likely that if germ-line gene therapy were permitted, it would be used inappropriately; that is, there would be an inclination to employ it even when a comparable outcome could be accomplished using a less risky method. There is already concern that germ-line gene therapy advocates will make claims that the risks are less than they really are and the benefits are greater than will be realized. In 1992 and 1993, a scientific oversight panel of the NIH voted to withhold funding for a contract to support the gene therapy work of Dr. Steven Rosenberg, an aggressive and highly respected cancer researcher. Rosenberg, it is alleged, continued to conduct gene-transfer–based cancer therapy despite clear evidence that critical elements of the protocol failed to work as expected in the two years during which the initial clinical trials took place. Although there is no allegation of misconduct, issues of "questionable practice" have been discussed. These include neglecting the results of others, presenting sketchy data at hearings, continuing trials that had not worked as hoped. In general, the NIH oversight panel was concerned that the original clinical trials were premature, based on the available preclinical and preliminary data, and that the trials were better portrayed as research than therapy. In defense, Dr. French Anderson, another former NIH investigator, contends that, although the technology is not perfect, it is state of the art and critically needed to treat terminal cancer patients who have few other options. Focusing on the successes achieved in treating adenosine deaminase (ADA) deficient patients, gene therapists have gleaned enormous media success and financial support. In fact, recent reports indicate that many of the NIH-based gene therapists who developed the protocols of gene therapy are seeking and finding greener pastures in private and commercial gene therapy centers.

If the scientists, administrators, and venture capitalists involved in applying and commercializing gene therapy were appropriately thoughtful, there would be much less reluctance on our part to advocate the development and application of germ-line gene therapy for those few cases in which it is the therapy of choice. However, based on the real world risks that were just described and on the discordance of opinion relative to the ongoing trials of somatic cell gene therapy, we find insufficient potential benefit to justify such advocacy. Until we have almost certain knowledge of the real risks and benefits associated with germ-line gene therapy, we believe that the potential risks to all of the future descendants of the patient outweigh any benefit to a very small number of persons who might benefit. In the event of an unanticipated harmful outcome of a germ-line gene transfer procedure using mice or corn, the transgenic organisms can be killed, but clearly this option cannot be used with people.

To claim that our experience with nonhuman germ-line gene therapy provides us with a clear understanding of the potential risks involved in human germ-line gene therapy is presumptuous and risky. Human beings are unique, and we are only beginning to discover facets of that uniqueness in terms of basic genetic phenomena. For example, we now know of five human genetic disorders that are based on mutations involving expandable and contractable trinucleotide repeats. This baffling and novel mechanism for producing mutation is totally unpredicted, and there is currently no explanation for its cause. Similarly, geneticists have only recently discovered another novel and unpredicted phenomenon, genetic imprinting. For a small but significant fraction of genes, in human beings and other species, the expression of the gene during early embryonic development varies according to its paternal or maternal origin. The biological role of imprinting and the molecular mechanism responsible for selective gene expression remain mysteries. But the effect of genetic imprinting and trinucleotide expansion, in terms of carrying out germ-line gene therapy, may be critical. Problems may not be discovered until the second or third generation.

Our argument is not that germ-line gene therapy should never be developed. It is that it should not be developed at this time. As pointed out in the previous paragraph, we have just recently discovered two completely unexpected features of genes. It seems likely that other such novel and bizarre genetic phenomena will be discovered over time. If the incentive structure of science were different than it now is, one would expect that scientists themselves would support a moratorium on the development of germ-line gene therapy in human beings. Indeed, the overwhelming number of scientists with whom we have talked support such a moratorium. However, it takes only a few scientists who have convinced themselves that they know that the risks are only imaginary and that the benefits are real for germ-line gene therapy to become a field in which scientists compete to be first. It is to prevent those few arrogant scientists, motivated often by support from profit conscious venture capitalists, from initiating such a competition that we urge the vast majority of scientists to support a continuing moratorium on the development of germ-line gene therapy in human beings.

It is not that people are incapable of successfully using powerful technologies, although existing problems with nuclear waste, toxic dump sites, and acid rain indicate that this concern is not wholly without merit. Nor do we contend that people should be prevented from shaping the future of their own species, although one can certainly and easily imagine inappropriate uses for gene therapy for either negative or positive eugenics. Rather we hold that this or any technology that poses even a small possibility of causing great harm to many people cannot be justifiably used to provide benefits to only a few, even if those benefits are great. The presence of alternative, less risky procedures is, as pointed out in Chapter 2, a morally relevant feature of the situation.

Our key concern is the bias in favor of scientific progress among those who are and will be involved in developing gene therapy protocols and in evaluating their successes and failures. The promise of national and international recognition, of prizes, awards, patents, and grants, of all measures of status, wealth, and power is a potent incentive to overstate successes and benefits, to take unacceptable risks, and to dismiss valid objections. The loyalty of scientists to one another and their reluctance to interfere with any research project that their scientific colleagues wish to pursue make it very likely that some mistakes will be made. In cases where there are a great number of people who may be put at significant risk, all impartial rational persons would agree that caution must prevail.

The ethical problems raised by germ-line gene therapy are far more complex than those we have discussed in the preceding chapters. Even the question of abortion, which we consider to be unresolvable, is one in which we know what the issues are and why they cannot be resolved. We do know enough about germ-line gene therapy to be confident that it is still too early to lift the moratorium on its use. However, as techniques become more sophisticated and as we can replace or repair defective genes with greater precision, problems that now seem speculative or that have not even been considered may arise. We regard this application of morality to the new genetics to be just the beginning of an ongoing project that should take on added importance with each new advance in genetic knowledge and technology.

APPENDIX

Principles of Human Genetics

A Typical Human Cell

A typical human cell is shown in Figure 1 with its nucleus containing the forty-six strands (twenty-three pairs) of DNA and their associated proteins and other cofactors that make up the chromosomes (shown enlarged in the middle box). Each chromosome is but a single coiled strand of two molecules of DNA (the double helix). These are drawn at the bottom of Figure 1 as two horizontal parallel strands of alternating sugar and phosphate components, each sugar having an attached base, A, T, C, or G. As discussed in the text, A and T have a specific binding affinity for each other and likewise G and C, and this is what holds the two strands of sugar-phosphates together. These molecules are about 250 million bases long in the largest chromosomes and 50 million in the smallest.

Meiosis

The twenty-two pairs of nonsex chromosomes (autosomes) of one human parent are shown in Figure 2; at the far right are an XY, should the parent be male, and an XX, should the parent be female. The sperm or egg contains but one of each parental pair and which one of the pair (grandfather's or grandmother's) is purely at random. Thus there are 2^{23} possible combinations in the formation of the egg or sperm. Also shown for completeness (at the far right) is a single mitochondrion (M); mitochondria are the energy-producing organelles inside each cell. Each contains a small single circular chromosome of about 20,000 bases. Only the egg transfers this amount of genetic information (and a few very rare diseases), resulting in another type of inheritance passed directly through the maternal line. Not shown, however, is *recombination*, or *crossing-over*, which is the process whereby a symmetrical exchange occurs between a pair of homologous chromosomes (see Figure 3). This occurs about three dozen times in the formation of an egg or sperm, approximately at two or three places on the big chromosomes and once or even not at all on the smaller ones. Thus each chromosome in a grandchild ends up as a mosaic of the four grandparents, emphasizing the degree to which evolution goes to create genetic heterogeneity for survival.

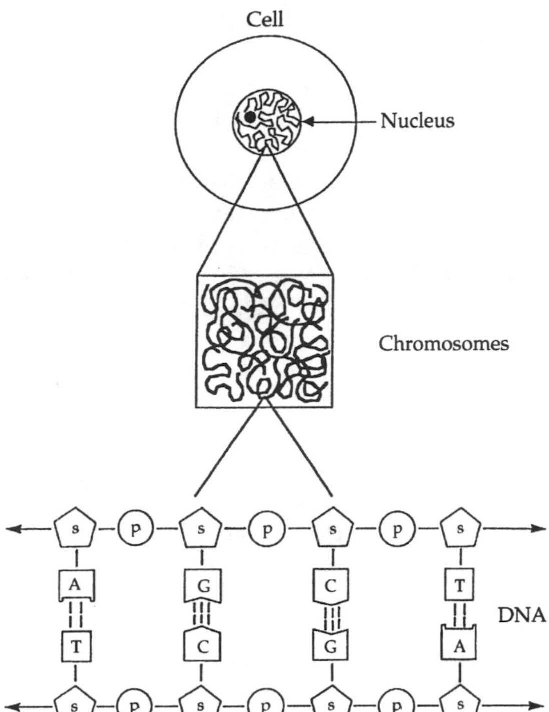

Figure 1 Typical human cell

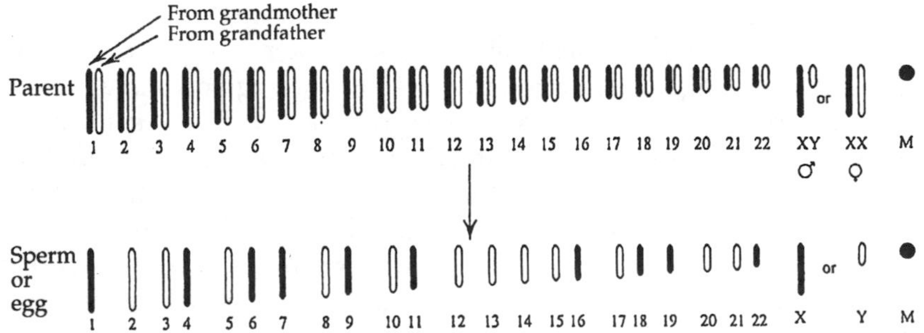

Figure 2 Meiosis (formation of egg or sperm)

Recombination

Four identifiable markers have been mapped to the chromosome illustrated in Figure 3, labeled A, B, C, and D at four positions (loci). Each marker has two or more variations (alleles), meaning that they can be chemically told apart; that is, *A* can be distinguished from *a*, *B* from *b*, *C* from *c*, and *D* from *d*. In the diagram, a recombination (crossing over) has occurred in the formation of the egg or sperm in the course of meiosis, as recognized by analysis of subsequent offspring. If this recombination between *C* and *D* occurs once per 100 offspring, *C* and *D* are said to be one centimorgan (cm) apart. One centimorgan is about 1 million bases in physical length, but in "hot spots" for recombination may be 100,000 bases or less; in "cold spots," recombination loci may be well over 1 million bases. More confounding, at any given location, the recombination rates differ between the formation of eggs or sperm. Thus the Mendelian, or genetic, map varies between males and females owing to different frequencies of recombination, whereas the physical map, meaning the number of bases between two loci, is more or less identical.

Linkage

M is a sequence of DNA at a specific locus, which is highly variable from person to person in a given population (highly "polymorphic" or highly "informative" in genetic lingo, with multiple alleles). In a given pedigree, if Huntington disease is found in affected individuals with one variant (allele) of the marker—namely *M4, HD* and *M* are probably "linked" closely together on the same chromosome. If a fetus in that pedigree has a parent with *HD* and the fetus also bears marker *M4,* there is a high probability the child also has the *HD* gene. If, in a large pedigree, *HD* and a marker for *M* appear together 95% of the time, they are said to be 5 centimorgans apart, if 98% of the time 2 centimorgans, and so forth. From this, one can easily deduce that the aim of genetic mapping is to find informative (polymorphic) markers along all twenty-three chromosomes and also along the DNA in the mitochondria, meaning some 3,000 markers to achieve a map saturated at 1 centimorgan degree. With this degree of saturation, errors in diagnosis by linkage analysis will be 1% or less.

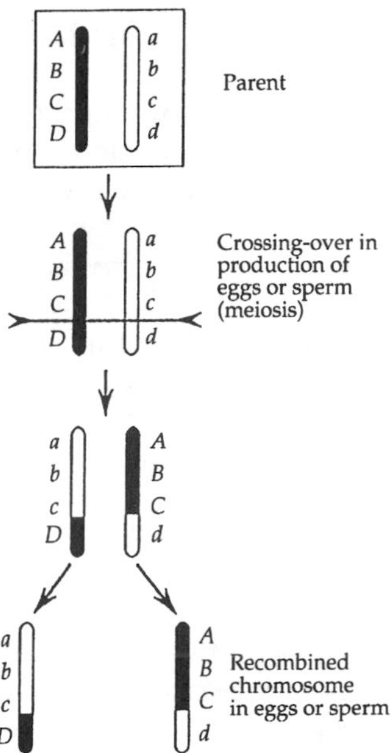

Figure 3 Recombination (reshuffling of genetic information into eggs and

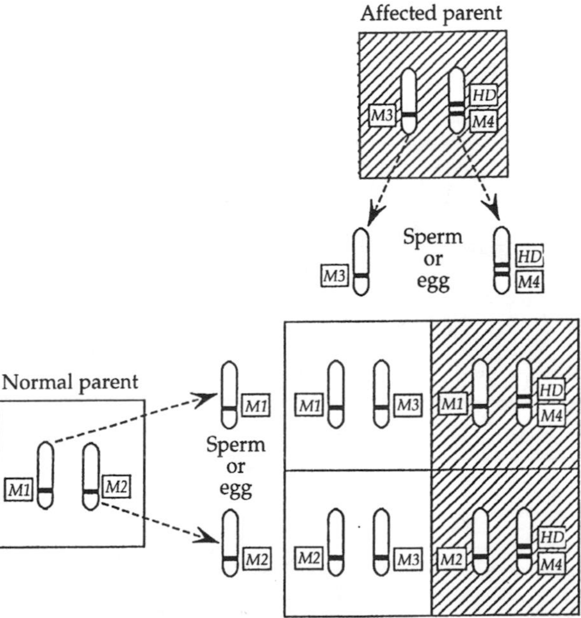

Figure 4 Linkage (genetic information close together travels close together)

Patterns of Inheritance

The schemes in Figures 5 through 8 show one pair of homologous chromosomes from one parent in the upper box and those from the other parent in the box at the left. As hypothesized by Mendel, the gamete (sperm or egg) contains only one of the two and which one of the two is selected purely at random (see Figure 2). Thus, from each parental pair of chromosomes, there are two types of sperm and two types of egg, and, after fertilization, there are four possible offspring, shown as the four squares. In dominant inheritance (Figure 5), one copy of the gene is able to express the inherited trait (diagonally hatched box). Because one-half of the sperm or one-half of the eggs from the male or from the female parent with the altered gene, respectively, are affected, the risk of any offspring bearing the trait (diagonally hatched box) is one in two.

In recessive inheritance (Figure 6), both parents are phenotypically normal but, as carriers (shaded boxes), are genotypically abnormal. One-half of their eggs and sperm, respectively, will carry the abnormal trait and the probability is therefore 1 in 4 that a child will receive both altered genes (diagonally hatched box) and 1 in 2 that a child will receive one copy and will be a carrier like his or her parents.

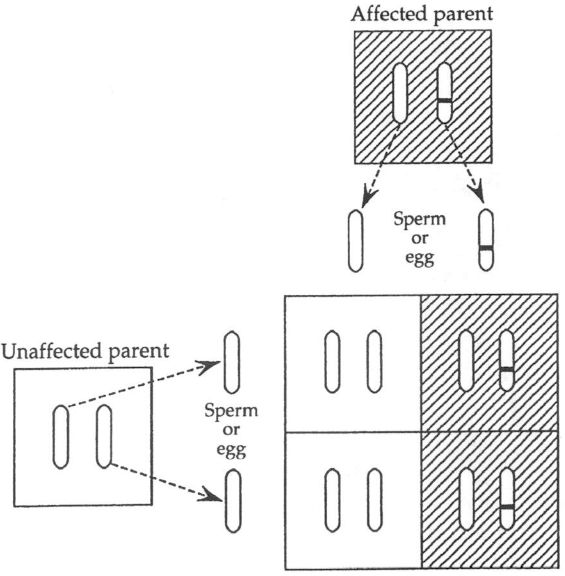

Figure 5 Dominant inheritance

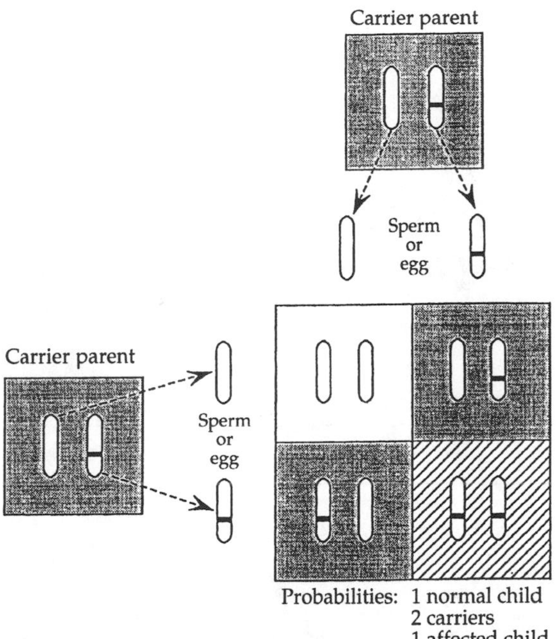

Figure 6 Recessive inheritance

In sex-linked inheritance (Figure 7), the trait is carried on an X chromosome. The carrier mother (shaded box), bearing the two X's of the female, is phenotypically normal, but half her eggs will carry the trait, meaning that one-half of her sons, being males with but a single X chromosome, will show the genotype (diagonally hatched box). Of the daughters, one-half will be genotypic carriers (shaded box) like the mother, but all will be phenotypically normal.

In recessive inheritance acting like dominant inheritance (Figure 8), one parent is lacking a gene that, as a recessive gene, is not phenotypically expressed (shaded box). Thus two of the four possible offspring will inherit this potentially defective trait. If, however, in either the carrier parent or in one of the carrier offspring there is a mutation in the normal allele inherited from the normal parent, that cell in which both copies of the gene are then lacking will behave abnormally as in classical recessive inheritance (Figure 6). Should the gene be a tumor-suppresser gene, that single cell becomes cancerous and will grow and divide, now unchecked in contrast to normal, noncancerous cells. This sequence is what happens in familial cancers such as retinoblastoma, familial breast cancer and some cancers of the bowel. As more studies are reported, more malignancies caused by this genetic event are being found.

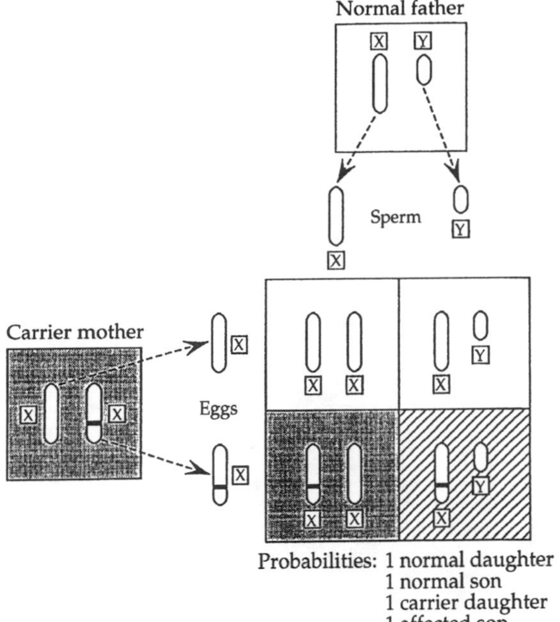

Figure 7 Sex-linked inheritance

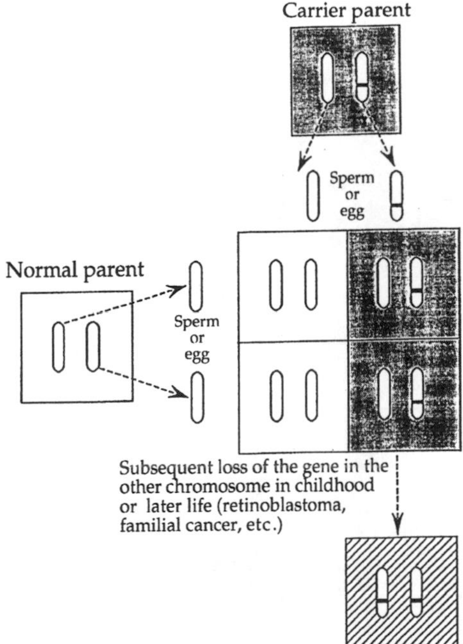

Figure 8 Recessive inheritance acting like dominant inheritance (as in tumor-suppressing genes)

Index